For Rafer

with best wishes

J. + Juliana Christy

ACHIEVING
THE RARE

Robert F. Christy's Journey in Physics and Beyond

ACHIEVING
THE RARE

Robert F. Christy's Journey in Physics and Beyond

I.-JULIANA CHRISTY
California Institute of Technology, USA

World Scientific

NEW JERSEY · LONDON · SINGAPORE · BEIJING · SHANGHAI · HONG KONG · TAIPEI · CHENNAI

Published by

World Scientific Publishing Co. Pte. Ltd.

5 Toh Tuck Link, Singapore 596224

USA office: 27 Warren Street, Suite 401-402, Hackensack, NJ 07601

UK office: 57 Shelton Street, Covent Garden, London WC2H 9HE

Library of Congress Cataloging-in-Publication Data
Christy, I.-Juliana.
  Achieving the rare : Robert F. Christy's journey in physics and beyond / I.-Juliana Christy.
  pages cm
  Includes index.
  ISBN 978-9814460248 (paperback : alkaline paper)
  1. Christy, Robert F. 2. Physicists--United States--Biography. 3. Nuclear physics--United States--History--
20th century. 4. Astrophysics--United States--History--20th century. 5. California Institute of Technology--
Biography. I. Title.
  QC16.C52C47 2013
  530.092--dc23
  [B]

                                                    2013012736

British Library Cataloguing-in-Publication Data
A catalogue record for this book is available from the British Library.

Printed in Singapore by Mainland Press Pte Ltd.

Preface

This book arose in an unusual and unexpected way.

Ten years ago, in 2003, I was informed of the presence of invasive cancer. It had started with breast cancer but had spread into the lymph system, and imaging showed an invasion into the liver and kidney. Until then I had worked full-time as an astrophysicist at the California Institute of Technology (Caltech). However, I now decided to devote most of my time and energy to my fight against the cancer. Additional motivation for this decision came from the fact that most of my close relatives from earlier generations had died of cancer.

I was touched by the support given to me by my boss Prof. Tom Tombrello, Chairman of the Division of Physics, Mathematics, and Astronomy at Caltech. He gave me permission to take an indefinite leave of absence to fight the cancer while keeping my academic position for me. This was of crucial importance to keep up my confidence in myself and in what I was doing. I had a great deal of support from many doctors and from my family. My strongest support came from my husband, Robert F. Christy — he often doubted my unconventional methods, but he believed in me and in my ability to find a solution.

When I returned cancer-free after several years I approached the new division Chairman, Prof. Andrew Lange, and asked him whether he wanted me to go back into astrophysics or do something totally new, such as writing a book of my recollections of the "Giants of Caltech." I was in a unique position to do the latter: I had many fascinating personal recollections of the eminent professors and trustees of Caltech, mostly from the decade when my husband was Provost (during part of which he was also the Interim President). Prof. Lange answered, "Write the book."

The first chapter was the one to be on my husband, Prof. Robert F. Christy, but I soon found that I had far more material than would fit into a single chapter. His story ended up as an entire book — this one.

I wrote much of the first draft of the book in May and June of 2011 while I was recovering from a major elective surgery. At that time, my husband was still available to answer my many questions. During the next year, while my health improved, his declined. I was trying to help him by using some of the unconventional techniques that had worked so well for me, and there were times when it looked promising. He had always been a survivor, but this time it was not to be. He passed on, smiling, on Wednesday October 3, 2012 in our master bedroom at home; my hands were in his while I kissed him on the right cheek and the forehead, and our older daughter Ilia kissed him on the left cheek.

The final revisions for this book, including the addition of Robert's last year and of memories contributed at his funeral, were completed in early 2013.

I was always taken by the way that Ralph Leighton wrote his books on Dick Feynman. They were written in the way Dick Feynman would speak. One can almost hear his voice, with his Brooklyn accent.

My husband also had a unique manner of speaking, always grammatically correct and beautifully rhythmic. He spoke in a deep, sonorous voice. He loved poetry and often quoted it in daily life. To me, his way of speaking and writing resonated like poetry. He also had a unique and subtle sense of humor. I could not reproduce his way of speaking via my own writing; therefore I included direct quotes from him as often as I could. Several interviews with Robert Christy had been recorded and transcribed, and extensive portions of these have been included at appropriate places in the text.

Especially revealing are three interviews by his own family. The first took place in 2001 at the family's ranch, where he was questioned by our two daughters Ilia and Alexa as well as by myself.

The second family interview was in 2006, when our daughter Ilia and her husband Chris Wakeham organized a week-long family get-together in Santa Fe. Robert's son Peter Christy and Peter's wife Heidi Mason organized excursions to Los Alamos and to its beautiful surrounding area. While at Santa Fe, Robert was interviewed at length by his sons Ted and Peter and their respective wives Vera Nicholas and Heidi Mason, as well as by our daughter Ilia, her husband Chris, our daughter Alexa, and myself. His three young grandchildren Liana, Tess, and Jack Christy also quizzed him. Many of the questions were directed towards Robert's experiences at Los Alamos while on the Manhattan Project. The lengthy job of transcribing this five and a half hour interview was accomplished by Dr. Arnold I. Boothroyd.

The third family interview was in February 2011. While recuperating in a hospital Robert recounted his memories of his sons Ted and Peter to Antonella Vigorito, who had been an administrative aide in the Christy home for several years.

With the permission of the California Institute of Technology Archives (the copyright holder), a number of quotes were included from Sara Lippincott's 2006 interview that was published in *Physics in Perspective* **8**, 408-450. (Note that this interview was also published by the Caltech Archives in a slightly different form on the World-Wide Web at http://oralhistories.library.caltech.edu/129/ or http://resolver.caltech.edu/CaltechOH:OH_Christy_R.)

While writing the book I often had questions about details of Robert's earlier life. I would circulate between my study and his in our home, pose the questions to him, and immediately write down his answers in his own words. Some of these answers are included as direct quotes.

A large number of quotes are included from letters that Robert wrote to me and to others. They display his love of life, his values, and his integrity.

I kept a diary during some periods. This served as an additional resource for details from those years.

In writing this book, I had tremendous support from our colleagues. Prof. Rudy Marcus asked me to include more of Robert's administrative accomplishments, and made other suggestions that helped to improve the book. The present title I owe primarily to Prof. Kip Thorne, who also suggested improvements for the organization of the chapters. I am grateful to Prof. Freeman Dyson who was so supportive of this biography. I wish to thank Prof. Robbie Vogt for all of his encouragement. I am indebted to both Prof. David Politzer and Prof. Rudy Marcus for their positive comments on the book when they acted as reviewers for the publisher. Professors Charlie Campbell and Ward Whaling suggested several significant additions and corrections. I also wish to thank Prof. Jack Roberts, Prof. Francis Clauser, Dr. Judith Goodstein, Mrs. Francis Yariv, Mrs. Marylou Whaling, and Mr. Frank Rubin for their helpful and supportive comments.

Prof. G. J. ("Jerry") Wasserburg contributed the tale of Robert Christy's necktie (see chapter 9). Charlotte ("Shelly") Erwin of the Caltech Archives permitted us to photograph the framed necktie itself.

I am indebted to Mrs. Rose Bethe, who introduced me to the World Scientific Publishing Company as well as providing a great deal of encouragement.

I am very grateful to our daughters Ilia and Alexa Christy who read the manuscript with great care. They made suggestions and contributed details, anecdotes, and quotes. Ilia made a tremendous effort that improved the manuscript in many ways. She was particularly helpful in pointing out places where I had succumbed to my tendency towards long, convoluted, Germanic sentences.

I am especially indebted to Gail Christy, who gave me many of the old documents and photographs of the Christy family.

I also wish to thank Antonella Vigorito whose cheerful greeting to my husband "Good Morning, Professor Christy," along with her elegant appearance, tended to get a smile out of him. She was the one who interviewed Robert in the hospital in 2011, when he recounted his memories of his sons Ted and Peter.

In early 2012, at age 95, my husband Robert decided that he no longer wished to be taken to the hospital for every ailment, so our daughter Ilia found a first-class hospice service, Seasons Hospice, who would assist us in our home with his medical problems. This allowed him to remain in our home until the end of his life. They were responsive, sensitive, and very professional: the best that one could have imagined. I would also like to thank Prof. Jack Roberts for recommending an outstanding physical therapist, Will Locklear, who came to our home three times a week in 2012 and was beginning to make progress in fighting the atrophy of Robert's muscles.

It was a pleasure to work with our editors at World Scientific Publishing Company. Jessica Fricchione Barrows, the Commissioning Editor in New Jersey, could not have been more encouraging, helpful, and prompt; her suggestions greatly improved the book. Alvin Chong and Jimmy Low, the Production Editor and Artist from Singapore, were also most helpful and prompt in answering our questions about the more technical issues of manuscript preparation as well as the book's cover design.

Without Dr. Arnold I. Boothroyd, this biography of Robert F. Christy would never have been written. He did much of the research, the scanning and preparation of photos and documents from our files, and the actual typing and layout of the text and photos. Arnold did a tremendous job in getting all the photos integrated into the text.

Most of the photographs, letters, and newspaper clippings were found in Robert's files; many of the rest were from my own files. Unattributed photos are Christy family photos. "Caltech photos" are included with permission of the

California Institute of Technology (Caltech), in particular the Caltech Archives. "RERF photos" are included with permission of the Radiation Effects Research Foundation. "LANL photos" are included with permission of the Los Alamos National Laboratories. Photos and documents from the Atomic Energy Commission (AEC) are in the public domain, as we were kindly informed by the Department of Energy. An image of Gary Sheahan's painting "Birth of the Atomic Age" is included with the permission of the Chicago History Museum. Other photos are used with permission of their respective copyright holders. We are grateful to Dr. Joyce Guzik for supplying us with photographs from the last scientific meeting that Robert was able to participate in, the Stellar Pulsation Meeting of 2009 in Santa Fe, New Mexico. Articles from the Pasadena Star-News, the Los Angeles Times, the Caltech News, and the California Tech are included with the kind permission of these publications. Letters to Robert Christy are quoted with permission of the senders or their heirs.

I.-Juliana Christy

Contents

Chapter 1

His Childhood

Robert F. Christy was born on Sunday morning, May 14, 1916 in Vancouver, British Columbia, Canada. He was named Robert Frederick Cohen at birth, but his father changed the family name to Christy when Robert was two years old.

Robert's father was Moise Jacques Cohen, an immigrant from England who was 31 years old when Robert was born. He had come to Canada in 1908 to study Electrical Engineering at McGill University in Montreal. Robert's mother was Hattie Alberta Mackay, a native of Vancouver who had travelled across Canada to attend McGill, to study to become a teacher. She and Moise met at McGill. She was one year younger than he.

Moise and Hattie appear to have met in 1908, their first year at McGill, and to have very soon become committed to each other. Moise obtained two 14 K gold lockets, one to hold two miniature photos of Hattie, and one to hold a lock of her hair. His initials "MJC" and the year "1908" were engraved on the outside of each locket.

Robert's father Moise was clearly not a penniless immigrant to Canada. Not only did

Moise Jacques Cohen's locket (engraved with MJC and the year 1908) holding photos of Hattie Alberta Mackay

he have the means to buy these
gold lockets, but he also was
able to travel to England to
attend his sister's wedding in
September 1910, when he was
half-way through his four years
of undergraduate studies at
McGill.

Moise and Hattie were
married on September 21,
1912, not long after they graduated from McGill. Shortly after
their marriage they moved to
Vancouver. Robert's older
brother and only sibling John
Christy Cohen was born there
the following summer. Robert
was born three years later.

Moise's locket holding a lock of Hattie's hair; the back (top right) is engraved "MJC 1908"

The Origin of Robert's Names

Robert's first name was taken from his maternal great-uncle Robert Wood, who
was well-liked because of his sunny personality even though all his business
attempts ended up as failures. Robert's middle name "Frederick" came from
his paternal side, from the second husband of his paternal grandmother Alice
Frances Cohen (nee Jones). She was widowed at age 22. In 1902 when she
was 35 years old, she married Frederick Alexander Christy. Even though
Frederick Christy died less than a year later, he had a long-lasting impact on the
whole family. When Alice Cohen had first been widowed early in life, she first
worked as a grocer to support her three toddlers, aged 1 to 4 years old. She had
later worked as a nurse far from her children, who were teenagers by that time.
But after her marriage to Frederick, Alice suddenly appeared to be financially
well off. She was able to embark on a year-long round-the-world trip. She was
able to purchase second-class tickets on a trans-Atlantic luxury liner, the
Titanic. (A passenger list recovered on a steward's body lists the Christy
family as travelling first class, but this list must have been in error — or

perhaps outdated: it is possible that Alice initially chose first class tickets, but later reconsidered and exchanged them for the less pricey second class tickets.) At her death in 1939 Alice left a substantial estate of £33,000, despite this being at the end of the Great Depression.

Alice's children were from her first marriage, but they too held her second husband Frederick Christy in great esteem and affection. Her two daughters, Rachel Julie and Amy Frances, took on the surname Christy rather than Cohen sometime prior to their journey on the Titanic in 1912. Alice's son Moise, Robert's father, gave his first son the middle name Christy

McGill graduation photo of Robert's father Moise Jacques Cohen (who later changed his surname to Christy)

and his second son the middle name Frederick. Moise changed his own last name from Cohen to Christy in August 1918. His sons then became John Christy Christy and Robert Frederick Christy.

The esteem and affection felt by the family towards Frederick Christy was also evident in that Alice's middle child, Rachel Julie, chose to be buried with him when she died unmarried in 1931 at age 44. Alice herself also chose to be buried with him when she died in 1939 at age 72.

The Sinking of the Titanic

Robert's grandmother Alice and her two daughters (Moise's younger sisters Rachel Julie and Amy Frances), along with Amy's new husband Sidney Samuel Jacobsohn, embarked on the Titanic in April 1912. They planned to attend

Moise's graduation as an electrical engineer from McGill University and his marriage to his fiancée, Hattie Alberta Mackay. (Hattie would likewise be graduating from McGill, as a teacher.) Afterwards the English family planned to travel across Canada to meet Hattie's family in Vancouver. They would continue on around the world by ship, visiting relatives in Australia and South Africa before returning to England.

The Titanic sank on the night of April 14, and Amy's husband Sidney Jacobsohn died when he went down with the ship. Alice and her daughters Amy and Julie spent several hours in a lifeboat before being picked up by a ship called the Carpathia. They were taken to New York, where Alice gave a newspaper interview. Moise and Hattie's wedding was postponed, but Alice and her daughters travelled on to Montreal to meet them, staying for three weeks. They cancelled their plans to meet Hattie's family in Vancouver, and returned directly to England rather than continuing on with the rest of their around-the-world trip.

Marriage of Moise and Hattie (Robert's Parents)

Moise and Hattie were married half a year later on September 21, 1912 in the All Saints Church in Montreal. A detailed marriage contract was executed at that time which suggests that Hattie and any future children would be separated from the English estate. Hattie was promised a substantial amount of cash ($5000) to be given to her by Moise at the time of the marriaage. If Hattie predeceased Moise this money would revert to him rather than being part of her estate. It is not clear whether she actually received the entire sum or not, but she clearly received at least a large fraction of it judging by the luxury items in her possession at the time of her death. The marriage contract also stipulated that if Moise predeceased Hattie, another substantial cash sum ($10,000) would be paid to her from Moise's estate (anything over this would go to his children).

If on the other hand Hattie predeceased Moise, her heirs

Robert's mother Hattie Alberta Mackay, shortly before her marriage

Moise with his first son, Robert's older brother
John Christy Cohen ("Jack")

would have no claim to his family's
estate. In other words, if Hattie died her
Vancouver relatives would not be able to
claim any of this money, and Moise
would care financially for any children.

After the marriage Moise and Hattie
moved to Vancouver where Hattie's
relatives lived. The first Vancouver
address we have for them is from letters
dated February, 1916: Suite 43, 621
Seventh Ave. East, Vancouver, B.C.
However, this may have been Moise's
business address rather than their home.
Robert Christy remembers living in the
Kitsilano area when he was small, a
district in Vancouver a few miles west of
the above address. Kitsilano is a beautiful
area close to the ocean, beaches, and the
gorgeous Stanley Park. It is also not far
from the downtown business center of
Vancouver.

Moise and Hattie's first son, John
Christy Cohen ("Jack"), was born on Sunday evening August 3, 1913. Their
second son, Robert Frederick Cohen (called "Bobbie" in his early years) was
born three years later, on Sunday morning May 14, 1916.

Robert experienced his
father's presence only for the
first year of his life. Moise
joined the Air Force on
September 24, 1917, just over
a year after Robert's birth.
Moise served for one year and
was discharged on September
3, 1918, just before the war
ended on November 11. His
discharge papers state that
this was due to "his services

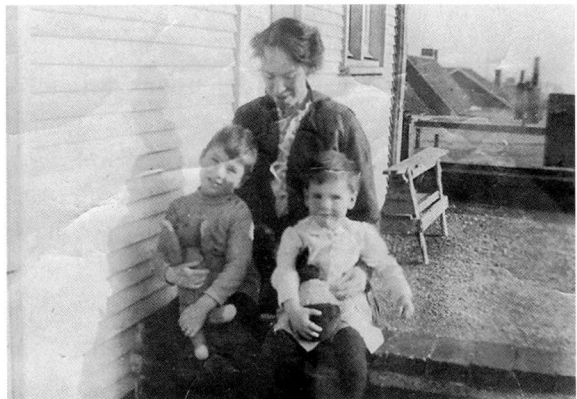

Robert with his mother and older brother

being no longer required upon annulment of transfer from R.F.C. to R.A.F." The Royal Flying Corps (R.F.C.) in Canada was a training organization for the Royal Air Force (R.A.F.) of Great Britain.

On August 31, 1918, just before leaving the Air Force, Moise changed his family's surname from Cohen to Christy by deed poll.

Robert at age 2, holding his father's pipe

Death of Robert's Father

After leaving the Air Force, Moise obtained a position as an electrical engineer with Ontario Hydro in Oshawa, Ontario. Hattie and their sons Jack and Robert remained in Vancouver, on the other side of the country. Only two months later tragedy struck the family: Moise died by electrocution while on the job on November 4, 1918. His death was "sudden" and "accidental" according to the death certificate. He was only 33 years old, and his widow Hattie was only 32; their two sons were aged 5 ¼ and 2 ½.

At this point, we have a Vancouver address for Hattie that is the same as that of Hattie's maternal grandmother Wood: 908 Broughton Street, Vancouver, B.C. This address was taken from a Power of Attorney dated December 11, 1918 that Hattie gave to her mother-in-law Alice Cohen Christy to handle Moise's financial affairs in England.

With Moise's death there was no source of income to support the young family. However there was a substantial life insurance payment of $4000 from a policy that Moise had taken out shortly after Robert's birth. In addition, there was cash from the sale of English War Bonds that Moise had owned (worth £550, i.e., about $2750). The power of attorney mentioned above was to enable Alice Christy to sell these bonds and send the funds to Hattie. These two cash sums added up to about two thirds of the $10,000 that had been promised in the marriage contract. There were small continuing payments from Workman's Compensation. Alice Christy also sent annual financial gifts from England to Hattie and the two children. Although Robert and his brother Jack were not entitled to any of the English estate, Robert remembered sending thank you letters to Alice and his two English aunts for their financial gifts each Christmas. Much later, when Robert turned 21, his English relatives sent a substantial amount of money to him and his brother — Robert remembered it as being $5000 each.

Robert's Early Years

As a single parent Robert's mother decided to give all her attention to her two young sons, rather than seeking employment as a teacher for which she had been trained. She had been the first of her family to attend a university. She used part of the cash payment to buy a beautiful home in an excellent neighborhood of Vancouver. There was an attractive school down the street, a nearby forest with a creek, and inviting playgrounds, tennis courts, and a nearby golf course. The address was 3125 West 39th Ave. in the Kerrisdale area, in the western part of Vancouver not far from the University of British Columbia and the ocean.

Robert enjoyed a happy childhood there, climbing the wide fir trees. He told stories of how he climbed to the top of the trees and slid down from one bending branch to the next until he reached the ground, never getting hurt. He also built forts in the forest across the street from his home. His adventurous streak had an early start.

Robert remembered being sent to Sunday School for many years. Although his father had some Jewish heritage, Robert did not learn any Jewish traditions.

In the summer Robert's mother would take her two young boys across the border into the United States to spend vacations every year at a farm near the ocean on Lummi Island. They spent wonderful weeks there playing with other children, with farm

Robert at age 4

animals and horses, and swimming and playing in the ocean. Robert remembered riding on the large working draft horses.

Robert started working to help supply income before the age of 10. One of his jobs was as a newspaper delivery boy in his own neighborhood. He told stories of developing techniques to throw the newspapers from his bike at full speed which was especially challenging on steep hills. Occasionally he crashed but he nonetheless continued to improve his throwing technique. Another story tells of how he would hang onto street cars to pull him along on his bike and speed him on his newspaper route — and save on pedaling. He also earned money at the local golf course, caddying for the wealthy patrons.

Robert at age 5

Robert had various memories of his extended family. He remembered being intrigued when his aunt Amy ended up living in Africa. He also remembered visiting his maternal great-grandmother Wood in her Broughton Street house in Vancouver. He noted that many of his relatives lived in that house, including the three children of his great-grandmother Wood: his maternal grandmother Alberta Mackay, his great-aunt Maud, and his great-uncle Bob (after whom he was named). He heard that the only one with a paying job was his great-aunt Maud, who worked as a secretary and helped support the various business ventures attempted by her brother Bob, all of which eventually failed (the ones that Robert recalled included a candy store and a drugstore). However, Robert remembered his great-uncle Bob as having a winning sunny personality, always happy and jolly.

No More Gambling

When he was quite young, Robert played marbles with the other boys. This game had unexpected consequences. One day when he was about seven years old, he played "Marbles for Keeps," which was unusual for him, and he lost all of the marbles that he owned at that time. After that, he decided not to gamble any more.

He also used to play a game involving chestnuts. To get these chestnuts the boys had to climb a neighbor's chestnut tree and pick them. However the neighbor did not like the boys climbing his chestnut tree and called the police. The police came and spoke sternly to all the little boys, saying that they were not to trespass on the neighbor's property, nor take the neighbor's chestnuts. This was Robert's first and only run-in with the law — except for speeding tickets later in life.

Death of Robert's Mother

Tragedy struck once again when Robert and his brother Jack discovered their mother dead in her bed one morning in 1926. She had just been sent home from the hospital after a goiter surgery, and may have died of a blood clot. Robert was 10 years old, and Jack was 12 ½. Now they were orphans.

Grandmother Mackay, Great-Aunt Maud, and Great-Uncle Bob decided to move into the boys' home to take care of them. Robert remembered fondly the Sunday dinners with his grandmother's roasts. She would cook a turkey for the holidays and Robert would watch and learn from her. He used her recipe for the rest of his life when preparing the holiday turkey feast, and taught it to his own children.

Early Interests

Robert was encouraged to work as diligently as possible both in school and in paying jobs after school. Right from the start he much preferred academics to his after-school jobs. He excelled in his school work and skipped two grades. This had both advantages and disadvantages. He was very bright intellectually, but socially a little disconnected from his classmates who were significantly older than he was. He had trouble making friends with them, unlike the boys he had played with in his earlier childhood before his mother had died. There was only one exception, a boy named Oliver Lacey whom he befriended in high school. As Robert Christy once stated in a letter written when he was 54 years old, "…I am basically quite shy. I was very much so when a young man, and much of it has worn off, but still some remains."

Robert also missed out totally on certain basic courses that were only taught in the years he had skipped. He never had a class in geography even though he was interested in it. He also felt that he never learned proper English writing skills. He excelled in spelling, mathematics, and science, but felt that he was unable to present his feelings and thoughts in English essay writing. (His skilled and expressive writing from later in his life suggests that either he was being too hard on himself, or else that he improved his writing skills as time went by.) Robert turned into somewhat of a loner. His teachers recognized his academic prowess, and pushed him into competitive exams, one after another. Due to the stress of always being pushed, he ended up with ulcers for a while at the age of 14.

High School ~ 1930

Robert in high school at age 14

Robert remembered being interested in science from an early age. In addition to enjoying his mathematics, physics, and chemistry classes, he and his brother learned to build mechanical and electrical apparatuses. For example, they built a crystal receiving set so they could listen to the radio as they had not had a radio prior to that. He recalls an occasion in his physics class in high school when he asked the teacher the explanation for the fact that when you were in the bathtub, and your big toe was sticking out of the water and the rest of the foot was in the water, the big toe seemed so much bigger than the rest of the foot. The teacher laughed, and said that it was refraction. This incident illustrates Robert's inherent curiosity into natural phenomena and understanding how things work.

At Magee High School, Robert found a mathematics teacher who was interested in him and supported his learning. In algebra class this teacher allowed Robert to sit by himself and work through all the problems in the book on his own, while the teacher taught the rest of the class the material that Robert already understood from his reading and his homework. Later this teacher organized a special class in trigonometry for Robert after school. This subject would normally not have been part of the high school curriculum.

There was also a physics teacher whose teaching style created and encouraged interest among his students. This teacher did not always answer questions directly, but would answer in such a way as to encourage the student to think about the problem. For example, at one point Robert told the teacher that he had

read about night glasses (binoculars that
are used on ships to see better at night),
and asked how they worked. The teacher
pointed to a pair of binoculars, and just
indicated with his fingers that the
objective lens was large. That was enough
to allow Robert to figure out that the way
the night glasses worked was to have a
larger objective lens to gather more light,
so that one could see in dimmer light than
would otherwise be possible. This teacher
supported Robert's interest in physics and
his inquisitive nature.

Robert also found the machine shop
and woodshop classes both appealing and
useful. He learned to use simple machine

Robert in high school at age 16

tools such as lathes and drills, and learned
how to do basic carpentry.

Academic Successes

Normally one could attend the University of British Columbia (UBC) after
finishing Grade 11. However because money was short at home, Robert's
teachers encouraged him to stay in high school to the end of Grade 12, thus
saving the first year of university tuition and entering into the second year of
university afterwards. He thus graduated from high school in 1932, at age 16.

It turned out that this decision had a major impact on his life. During that
year while he was attending Grade 12, the Province of British Columbia changed
their provincial testing procedure, no longer requiring essay writing in English
but using only multiple-choice questions. Thus any weakness in essay writing
would not impact his performance on the exams as he excelled at multiple-choice
tests. The result was that he had the highest exam score in the province and was
awarded the Governor-General's Gold Medal for British Columbia upon his
graduation from high school. This provided tuition for his undergraduate years at
UBC, which was crucial as his maternal family did not have the funds for his
tuition. This was in contrast to Robert's brother Jack who did not win such a

prize, and thus never attended university. Instead Jack worked in sales to help support the family.

Winning this top prize also put Robert into the position of being an academic star, which continued for the rest of his life due to his later accomplishments.

Another consequence of winning this prize was that Robert met Dagmar Elizabeth von Lieven, the second prize winner, at the award dinner. He was only 16, and she was 17. It was love at first sight, but they waited nine years before getting married because they both wanted to finish their education before starting a family.

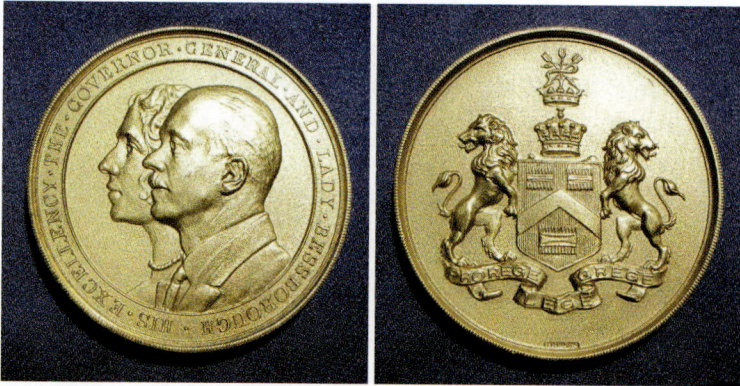

The Governor-General's Gold Medal for British Columbia that Robert was awarded on graduation from high school

Chapter 2

The Formative UBC Years

Robert Christy was fascinated by mathematics and by both theoretical and experimental physics. Chemistry, biology, and other fields of science did not have the same attraction. As soon as he completed the required courses in those fields, he specialized instead in mathematics and physics. As he said, "It became clear to me that my interest was more in physics than mathematics, so I went into, you might say, mathematical physics." He felt that the instruction in physics was more interesting although he also liked mathematics as a discipline.

Robert had entered the University of British Columbia (UBC) expecting to get a Bachelor's degree and become a teacher. While he was there he realized that it would make more sense to continue his education, especially since jobs were scarce in the 1930's due to the Great Depression.

Robert continued to excel during his undergraduate years at UBC. He liked some of the professors, but said that what probably helped most was that there were about half a dozen students, three in his year and two or three in the year ahead of him, who all had rather similar interests in science — mostly physics and mathematics. So there was a "core group" of students who were together a lot of the time and who provided competition for each other. There was also one physics teacher, Professor Hebb, who had a rather curious way of teaching. He would come into class and start to talk about some subject. Then he would pause and say "Well, now, let's see; how do we proceed from here?" as though he was trying to get his thoughts together. Whether this was actually the case or whether as seems more likely this was a device to encourage the students to think, it was very effective. One of

Robert Christy, while attending the University of British Columbia

13

the "core group" would generally suggest a way of proceeding. This participation of the students in the lecture was most effective in encouraging them to think about the subject and Robert enjoyed this teacher's classes.

The fact that Robert later went mostly into theoretical physics may have had a start in his UBC years. The classroom education was mostly theoretical, and focused on teaching the laws of physics. Additionally, none of the UBC faculty were doing any research in physics. Only Professor Shrum really had much contact with modern physics. He taught a course in atomic physics, which Robert found interesting. Robert believes that because nobody at UBC was doing research in physics, he lacked encouragement to proceed in experimental work.

Robert had two close male friends at UBC. George Volkoff was one year ahead of him in physics, part of the "core group" described above, and Robert had met him for the first time while at UBC. On the other hand, Oliver Lacey had been Robert's friend since high school.

Oliver chose to go into psychology at UBC and later became an alcoholic. Some years later, while Robert was far away at Berkeley, he learned that Oliver had died in a fire. Throughout his life Robert thought fondly of the friendship he had had with Oliver.

George Volkoff was an energetic go-getter who had a major impact on Robert's life. George spent a great deal of effort researching the best place to go for graduate work in theoretical physics. He discovered that J. Robert Oppenheimer's group at Berkeley was the leading group in North America. This influenced Robert to do his graduate work with Oppenheimer as well.

When Robert got his Bachelor's degree with First Class Honors in Mathematics and Physics in 1935, he did not think that he was prepared to apply for graduate work. However unemployment was very high and there were few jobs, due to the Great Depression. There was an opportunity at UBC for a few students to continue as graduate students while earning some money as teaching assistants. This provided a "free" education beyond the Bachelor's degree. Both George and Robert decided to take advantage of this, and stayed at UBC for another two years of graduate work in physics to obtain their Masters degrees and to increase their chances of being accepted for graduate work by Oppenheimer.

There were no professors at UBC who taught quantum mechanics, so a group of students organized themselves and essentially taught themselves an introductory course in that subject. Robert recalled, "The Master's thesis was a difficult issue because none of our professors were doing research. We had to essentially find a field of research to explore, and then do the research ourselves without any assistance. If there is no research going on, this is not an easy thing to do."

Robert first looked into the possibility of investigating aspects of the persistence of vision. This field was initiated by Hertz when he spun a rotating shutter in front of people's eyes to test at what speed the flickering appeared to cease and people perceived the intermittent images as continuous.

When he ran into a dead end on the vision project, at someone's suggestion Robert started an experimental project to measure the mobility of oxygen ions. These are oxygen atoms that have an extra electron attached to them. Since they are electrically charged, an electric field can cause them to move. As Robert described it:

> This involved constructing a complicated glass apparatus with various electrodes, in which the experiment was to be conducted, and this was my Nemesis. I learned how to blow glass, and fasten pieces together to make a vacuum apparatus, but the complicated piece of glass with all the electrodes in it had to be made by a professional. This was done, but at one point when I was beginning to work with it, it cracked and required repair, and I never succeeded in properly repairing it.

Robert received his Masters degree in 1937, but he did not consider that his experiment had had very satisfactory result.

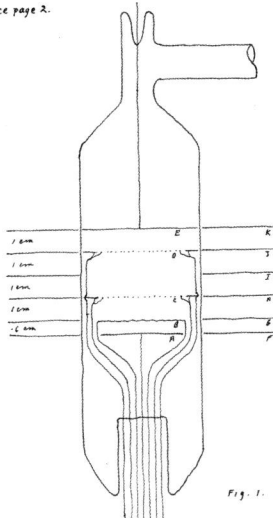

Diagram (left) and photo (right) of the tube that gave Robert so much trouble

Robert's Social Life at UBC

Robert had developed a close friendship with George Volkoff, who was a member of a distinguished group of émigrés who had left Russia when the Communists took over. George was dating Olga, another lively member of this group.

Dagmar von Lieven was also a member of this group, a quiet girl with a certain mystique around her. She was referred to as Princess Dagmar by the other émigrés, even though she tried to hide her aristocratic background (which is described in Chapter 4).

While George was dating Olga, Robert was dating Dagmar. When George and later Robert left for Berkeley, they stayed in contact via letters with Olga and Dagmar. Years later, after they all had finished their graduate training, George married Olga and Robert married Dagmar.

Robert had very fond memories of learning how to sail in the ocean around Vancouver while attending UBC. Dagmar's friend Jill Biller was an outgoing and lively young woman. She owned a sailboat, and Robert spent a considerable amount of time sailing with Jill and other friends, although Dagmar chose not to go with them. Many years later, after Robert had married Dagmar, Jill married a Mr. Simms. She was later

Robert swimming with his friends

widowed but was still alive in a seniors' facility as of 2010. Throughout her life, Jill tended to visit Robert (and Dagmar while she was alive). After becoming unable to travel due to her age, Jill continued to stay in contact with Robert by telephone.

Even at UBC itself there were some amusing incidents. Once during Robert's first year there was a meeting in the auditorium and a group of sophomores came in to disrupt it. Robert had stretched his long legs out into the aisle (he was nearly six and a half feet tall), and as the students came rushing down it they tripped over his legs and fell down in a heap.

Summer Jobs While at UBC

Robert's scholarship covered his tuition at UBC but not his living expenses. To help support himself and his maternal relatives, he found paying jobs during the summer vacations. These were varied and sometimes interesting.

One summer Robert's great-aunt Maud got him a job in her employers' office. The next summer she found him a job looking for gold in the middle of British Columbia. The miners used a big hose to divert water from a stream and washed away a tremendous amount of earth from the rocks, ruining a lot of beautiful countryside in the search for rocks containing gold. This search failed and no gold was found. Robert's task was to bring supplies to the mine, ferrying them in a small boat across a lake and then carrying them up to the mine. He did not enjoy hauling heavy barrels of nails up to the mine. He later said that this experience taught him that he would rather find a career which utilized his intellectual powers rather than his physical capabilities.

The next summer Robert got a job selling encyclopedias door to door. He clearly was not a salesman, and only managed to sell one encyclopedia in two weeks. That was his last attempt to work as a salesman. He spent the rest of that summer in the middle of British Columbia repairing railroads with a group of otherwise unemployed men. It was heavy physical labor. In the evening he was supposed to teach them English. However he found himself too exhausted from the heavy work he had been doing all day, and by the time evening came, he was too tired to teach.

Of all Robert's summer jobs, his favorite consisted of a job surveying a poorly mapped area in the middle of British Columbia. He hiked solo all day through the forests among lakes and mountains, establishing surveying marks. He also mapped the locations of the streams, rivers, and mountains, since the general terrain of this area was still unknown. He encountered grizzly bears and often had to find his way back to base camp in darkness. His initial salary was $90 per month, but

Robert on his surveying job in the interior of the province of British Columbia

due to his outstanding work this was increased to $120 per month. Deductions were taken from his salary of $1.25 a day for food supplied to the base camps, or about $38 a month, leaving him $52 the first month and $82 per month thereafter.

The surveying skills that Robert had learned proved useful again much later in his life. In the late 1980's and the 1990's, when he was in his 70's and 80's, he surveyed the area around the Spring Valley Ranch that he and his second family had acquired (see Chapter 17).

Robert always very much enjoyed hiking and walking in forests and mountains, even when nobody else was able to go with him. He had a special love of mountains which was expressed in many of his letters. In 1970 while flying from Washington to Los Angeles he wrote:

> We have just crossed the crest of the Rockies! The higher parts are already snow covered and the setting sun shining on the white peaks is a wonderful view. There is a hint of the "alpengluh" — have you seen it in its proper location? I saw it once in Obergürgl. … I love the mountains. The Sierras, the Rockies, the Alps — I have visited them all and climbed in all of them. I plan to do more.

Moving On from UBC

Robert's friend George Volkoff applied for graduate work with Oppenheimer and was accepted after completing his Master's degree at UBC. The following year Robert (who was a year behind George) also obtained his Master's degree, applied for graduate work with Oppenheimer, and was likewise accepted. As Robert put it, "George was accepted, and I guess he did well enough there that when I applied the following year, from the same institution and with a comparable record, I was accepted. In fact I was awarded a fellowship there for my first year at Berkeley." In the fall of 1937 Robert left Canada for good to go to Berkeley.

When Robert Christy turned 21 in May 1937 his paternal grandmother in England sent a substantial gift of cash to him, and an equal amount to his older brother Jack. Robert believes that it was $5000 to each of

Robert's older brother Jack and Jack's wife Jean, at their wedding

them. In the middle of the Great Depression this was a huge sum for a student to have, enough to live on for years. Robert decided to give *all* of his share of this cash gift to his brother so as not to feel guilty about going away and leaving Jack to take care of their older maternal relatives. Instead Robert lived on a shoestring budget at Berkeley: $640 in fellowship income for the first year, part-time jobs during the academic year, and full-time summer jobs. This shoestring budget often left him hungry, as described in the following chapter.

Chapter 3

Exciting Graduate Years with Oppenheimer

Robert Christy now joined a dynamic group of theoretical and experimental physicists. It was a very exciting time at Berkeley because the hot new subject in theoretical physics was nuclear physics, and Berkeley was the world center of nuclear physics. It was crucially important that Earnest Lawrence was also at Berkeley. He used his new cyclotron to provide experimental tests for the theories of nuclear physics, which later led to his winning the Nobel Prize. Robert recalled:

> Lawrence had his cyclotron, and he was in the center of experimental nuclear physics. It is said that he missed more discoveries because he was so involved in getting more and more intensity and higher energies out of his equipment that he didn't really pay much attention to the physics, so other people were getting the real discoveries in physics.

The group with whom Robert lived in 1937-38: Robert is on the left, and his friend George Volkoff is 3rd from the left

Robert (second from left at the back) among some of his fellow students

The California Institute of Technology (Caltech) was also a center of nuclear physics at that time, but not at the same level as Berkeley.

The dynamic group at Berkeley included Ed McMillan, a young professor who was busy building his synchrotron which likewise led to a Nobel Prize. There was Miss Chen-Shung Wu who also later obtained a Nobel Prize in experimental physics, and who greatly impressed Robert with her multifaceted personality. Decades later, when the question came up whether Robert's daughters should learn Chinese, his memory of Wu was a deciding factor for him.

The graduate students were also an impressive group. These included Eldred Nelson, Joe Keller, and of course, Robert's friend George Volkoff. The nine-month apartment rentals were always shared by several graduate students. Robert remembers one such period when he had to share a bed with George Volkoff, because there were too many students and too few beds. This is when he learned how

Performing a prank: George Volkoff is on the left, Robert is at lower right

to sleep on his side, carefully keeping to one side of the bed, while George carefully slept on his side on the other edge of the bed.

Oppenheimer's lectures were very difficult. Many students worked in pairs with one furiously taking notes while the other one listened, the two of them collaborating to help decipher the challenging material. Robert, however, always worked on his own. The lectures were full of content and were very difficult to follow.

To help with his expenses, Robert was a busboy one year in the International House. When hunger overcame him, he tried filling his stomach with milk, because milk was filling and inexpensive — however, he found that feeling of fullness

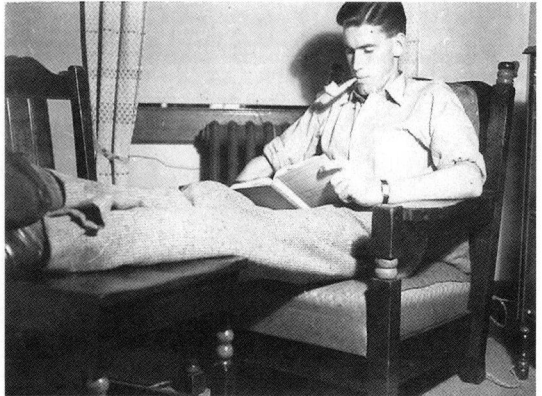

Above: Robert Christy as a graduate student at Berkeley (around 1940, about 23 years old).
Below: a remarkably similar photo of Robert's father Moise Jacques Cohen as a university student at McGill (around 1908, about 23 years old; note that Moise died in 1918, when Robert was only 2 years old).

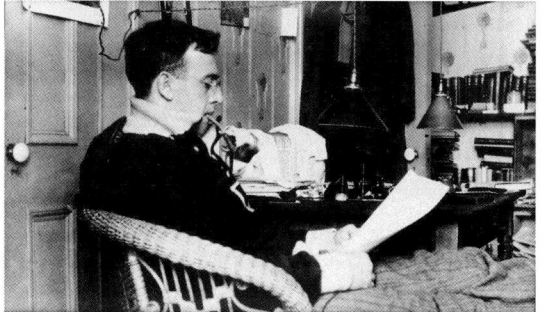

only lasted for an hour. He was known as the "three-rice-bowl man" at a local Chinese restaurant because he was always hungry and would fill up on cheap rice.

Some evenings Oppenheimer gave parties in his house to left-leaning groups to raise money from his guests for their causes. It was the time of the Great Depression and there was a great deal of unemployment and hardship. Oppenheimer hoped to help alleviate the hardships. Robert also attended these events sometimes, but of course he could contribute no money.

In addition to having a group of graduate students at Berkeley, Oppenheimer taught a course at Caltech each spring. He would teach at Berkeley for two semesters before going to Caltech for their third semester, which started in April.

When Oppenheimer moved to
Pasadena each April to teach at
Caltech, he would bring all of his
graduate students with him.

At Caltech, Oppenheimer
would stay in Professor Tolman's
guest house. The graduate students
would stay in various boarding
houses. Robert remembers these
boarding houses because the
landlady would provide meals in
addition to providing rooms to

Robert driving Oppenheimer's car, on a street near the
Tolman guest house where Oppenheimer lived while
he was teaching his spring-semester course at Caltech

house the students. This was in contrast to Berkeley where food was not provided with
the housing.

One summer, Oppenheimer asked Robert to drive his Cadillac convertible
from Berkeley to Pasadena. Oppenheimer and his new wife Kitty had chosen to
come to Caltech by train rather than in the open car because they had a newborn
baby.

Becoming Part of the Birth of Particle Physics

Cosmic rays are high-energy particles that enter the Earth's atmosphere from
outer space; some of them traverse the entire atmosphere to reach the Earth's
surface. Not much was known about them at the time. Today we know that they
are mostly protons with various other particles as well. At that time, it had just
been discovered that some of these cosmic ray particles were of intermediate
mass, roughly 200 electron masses (one tenth the mass of a proton). These
particles were called "mesotrons" at the time and are now called mesons.
Because of this discovery Robert's Ph.D. thesis focused on the new experimental
data.

Robert's thesis consisted of two papers on cosmic rays, one theoretical and
one a comparison with observations. Some of the theoretical calculations were
exceedingly complicated and Oppenheimer suggested that both Robert Christy
and Shuichi Kusaka (another student of Oppenheimer's) carry out the same
calculations independently, in order to ensure that they got it right. These
theoretical calculations were published in 1941 as a paper by R. F. Christy & S.
Kusaka, "The Interactions of γ-Rays with Mesotrons," in *Physical Review* **59**,

405-414. The "mesotrons" mentioned in the paper's title were highly-penetrating particles observed among the cosmic rays whose nature was unknown at the time (today we know that they are muons, also called mu mesons).

The second paper used these calculations, along with the experimental data, to attempt to place constraints on the properties of these "mesotrons." It was published immediately following the first paper in the journal as R. F. Christy & S. Kusaka, "Burst Production by Mesotrons," *Physical Review* **59**, 414-421. It was determined that the observed properties of the bursts were consistent with the "mesotron" being spin-0 or

Robert near the end of his graduate studies

spin-½ particle, but not with a spin-1 particle. It is now known that this mu meson has spin-½.

Robert's Years at Berkeley in His Own Words

In an interview with members of his family on November 18, 2001, Robert described his experiences at Berkeley:

I.-Juliana Christy (IJC): After you graduated from UBC [the University of British Columbia] with a Master's degree in 1937, what prompted you to go to Berkeley?

Robert F. Christy (RFC): Going to Berkeley was pretty straightforward because George Volkoff, whom I mentioned earlier, and who was a year ahead of me, had pioneered the way. He was interested in theoretical physics and so was I, and he had researched where the activity in theoretical physics was in the thirties. He found that the most active group in theoretical physics was in Berkeley under Robert Oppenheimer. He applied there — he may have applied elsewhere too, I am not sure — but he applied there with the hope of going down to Berkeley and studying under Robert Oppenheimer. He was accepted, and I guess he did well enough there that when I applied the following year from the same institution and with a comparable record, I was accepted. In fact I was awarded a fellowship there for my first year at Berkeley. My choice of Berkeley was essentially made because my friend George Volkoff had gone there the previous year and had found it a very satisfactory place to go.

IJC: How was the transition from UBC to someplace as lively as Berkeley? What was going on in daily life in your studies there?

RFC: Well, it was a very exciting transition of course. My living conditions at first were with a joint arrangement with some other students, most of them former UBC students. Those students were George Volkoff, Ken McKenzie, myself, Bob Cornog, and Ken McKenzie's wife Lynn McKenzie, who was a nurse. The five of us rented an apartment and lived together for a year. That made the living conditions easier in the transition to a new environment. The courses that I took there I found to be very challenging. I took courses in analytical mechanics and in electricity which were not unusually challenging, but the particular course that was a challenge was Oppenheimer's course in theoretical physics. This was known to be a very difficult course, so I had to work very hard that year.

Ilia Christy (IC): Can you speak a little bit about your graduate work at Berkeley and your colleagues at Berkeley?

RFC: I was immersed immediately at Berkeley in a very lively atmosphere of primarily experimental nuclear physics. There was a regular weekly seminar called the journal club. I think it met every Monday evening at 8:00 p.m. until 9:00 p.m., and various people would give brief reports of the order of ten minutes or so on all the current work that was going on in the field anywhere in the world as well as work at Berkeley. Since Berkeley was one of the liveliest places in nuclear physics because of Lawrence's presence with his cyclotron work and the large group that was working with him, this was a very exciting environment to be involved in. In addition there was a theoretical seminar in which there were presentations made on various theoretical subjects, both by advanced students and by post-docs who were working with Oppenheimer. I think that was a weekly affair. So the whole environment was one of being in the middle of a new field of science, namely nuclear physics and elementary particle physics. Some of my colleagues there I have already mentioned. George Volkoff was doing theoretical physics a year ahead of me with Oppenheimer. Ken McKenzie did experimental physics and ended up actually going to the department at UCLA and was there for many years. Bob Cornog was an engineer and I've had various contacts with him from time to time ever since, but I don't know precisely where he went. Another colleague who was a year ahead of me in theoretical physics was Phil Morrison [Phillip Morrison], who later became perhaps best known for his writings in Scientific American where he and his wife reviewed books and things for the Scientific American. But he had a very interesting career in physics in addition. He is now at MIT. A friend of his, also a year ahead of me, was Sid Dancoff [Sidney Dancoff], who was one of the few married graduate students at the time. Phil Morrison was married too, to his wife Emily. Sid was married to Martha, and they [Sid and Phil] did theoretical physics under Oppenheimer. Sid went to the University of Illinois just a year ahead of my going to Illinois Tech in Chicago. There were various other students I don't remember so well. There was Ed Gerjuoy who I've run into from time to time, but I don't know much about his career. There was Bob Peterson who eventually ended up, I think, at Sandia Labs or somewhere in Albuquerque. I've run into him there once or twice. There was Eldred Nelson who was, I think, a year behind me — I forget where he came from. Shuichi Kusaka came from UBC a year after I did and I collaborated with him later. There was another one of the year ahead of me whose name doesn't come to my mind immediately [Joe Keller], but I spent one summer working with him: Oppenheimer often had students do theses, and would have a junior student working with a more senior student as an assistant and to generally break them in. So I worked one summer doing very lengthy numerical calculations on the thesis of this student [Joe Keller], and the subject of that eventually formed a publication which I think my name was on jointly. It had to do with the determination of the value of the fine structure constant by comparing data on X-ray spin doublets with calculations. Later I did research with Shuichi Kusaka helping me on my thesis, and we have publications on that. Part of my thesis involved theoretical aspects of the interaction of the cosmic ray mesotrons (as they were called then — they are now called mu mesons) with matter, and the development of showers which under certain experimental conditions were called bursts. The other half of my thesis was on the examination of the experimental data on cosmic ray bursts in order to compare with the theory that we had worked out, about what the probability should be.

IJC: A key student of Oppenheimer was, of course, Serber. Where was he in this environment at Berkeley?

RFC: Bob Serber was a post-doc at the time. He was one of a number of post-docs who were working with Robert Oppenheimer for a number of years. Serber was there with his wife Charlotte, a very lively lady, and Serber got a job probably in 1939 or 1940 at the University of Illinois. Later on Leonard Schiff, who wrote a well-known book on quantum mechanics, came and was a post-doc working with Oppenheimer. Later still, Julian Schwinger was there as a post-doc. He later won a Nobel Prize for work in quantum electrodynamics jointly with

Tomonaga of Japan and Richard Feynman. Another one that I knew there was Willis Lamb. I think he may have received his degree with Oppenheimer perhaps about the time I arrived. He later went to Columbia University and won a Nobel Prize for his discovery of the so-called Lamb shift, which was a fundamental discovery in understanding the electromagnetic interactions.

IJC: Wasn't there also a woman in this group, a physicist?

RFC: Oh, there were various women in the group actually. There was a young woman probably a little bit older than most of us, Margaret Louis, and I really forget what her role was. She may have been an experimental student, not one of Oppenheimer's students, I think. There was another young Chinese woman, an experimental student that I knew there, Chen-Shung Wu, who later went to Columbia University and got a Nobel Prize for her work in beta decay. She taught us a game she called "Chinese badminton" in which you have a little feathered shuttlecock that you kept in the air by kicking with your heel in various ways. There was also a young lady who appeared there — I don't think she was a student — her name was, I believe, Muleika Barklay, and she was involved with one of the graduate students.

IJC: Was there considerable interaction in these theoretical seminars, and these Monday evening get-togethers, between all these various people through the years you were there?

RFC: These were indeed very lively affairs, in which there would be questioning from the audience and discussions and so forth. These were not just talks by someone, but they involved discussions with the audience, and it was quite common for the professor to ask pointed questions which would either lead to further elucidation or perhaps on occasion to embarrassment of the speaker, if the speaker did not properly understand the subject. They were very lively, active sessions.

IJC: Were these sessions a precursor of what was going on at Los Alamos later?

RFC: Not precisely, no. These sessions merely made people aware of each other and of what was going on in the field. For example, the spring of 1939 was a very lively period because that was when the discovery of fission became known in the U.S., and immediately people in all of the major laboratories devised experiments to verify the existence of fission and to proceed to learn more about what was going on. Many of those experiments were carried out by people that I knew and heard reports from immediately at Berkeley. For example, one of the experiments was the discovery of a new element, plutonium, which came from the decay of uranium-238 plus neutrons making neptunium-239, which beta-decayed into plutonium-239, which was fairly stable but which was alpha-radioactive. That was discovered by someone who later became a friend of mine, namely Ed McMillan, who was more senior (about ten years older), and Glenn Seaborg, who both later won Nobel Prizes and who were involved in the atomic energy program. All these discoveries were going on all the time there in Berkeley, and we heard about them all the time.

IJC: So things were not divided into theoretical groups and experimental groups, the way it tends to be today — there was a rich mixing of the two groups? And what about Lawrence's group?

RFC: Most of the experimental work that I mentioned was in Lawrence's cyclotron group, although I was aware of some of the chemistry work that was going on, but only as it impinged on the nuclear physics. The theoretical physicists were very interested in all of the experimental work that was going on there because that was providing the grist for their mill. The other was not true: the theoretical seminars were pretty well strictly confined to the theoretical physicists, but the experimental seminars were attended by everyone.

IJC: You just mentioned Ed McMillan. What was his position there at the time?

RFC: Ed McMillan was a young assistant professor in physics at the time there, and I think he was also participating in Lawrence's cyclotron work. I'm not quite sure of that. Lawrence was probably a full professor at the time, and I remember that Lawrence got involved with other work, and Ed McMillan had to teach his electricity course at one point.

IJC: What gave you your Ph.D. thesis subject; and did you work on any other projects? What were all these various projects?

RFC: I would guess that Oppenheimer must have suggested the thesis subject, but these subjects were chosen from things that were actively going on and where the students had interests. As I said before, it was a double-barrelled one. The mesotron at that time was a very strange cosmic ray particle which had the property of being very penetrating, not interacting very much with anything, and went down deep into the Earth. And yet the only theory that had been invented, by Yukawa, had been of a particle which would account for the forces in nuclei and would be very strongly interacting. So this was a puzzle, and there was considerable attention focused on the question of what was the nature of the mesotron, and did it have strong interactions or not? Since it was apparent that these cosmic ray particles did interact in the atmosphere occasionally to make bursts, that is, to make electromagnetic cascades, this might be a way of determining something about the particle. My job for my thesis was to calculate the interaction of mesotrons with nuclei via electromagnetic interactions, and find out whether this agreed with what was observed in cosmic rays. There were several different possibilities that were being considered for the mesotron. One was that it might have spin zero, and the other that it might have spin one. The thought was that it might be possible to determine which of these was true by comparing the calculated interactions with those observed in cosmic rays. With Shuichi Kusaka's help, I carried out these calculations. The calculations, by the way, for the spin one meson were exceedingly difficult. There were pages and pages and pages of complicated algebra, and it required two of us to do the calculations to make sure that they were done correctly. When these calculations were compared with the experiment, it was found that the spin one meson made many too many interactions and it was not possible. The spin zero meson seemed to be a possibility. But for good measure, I decided also to calculate what would happen if the meson were spin one-half, which no one expected at the time, and [when compared to the experiments] that also was a possibility. It turned out later that in fact the mu meson did have spin one-half, and this was a lucky chance, that I had calculated this ahead of time.

IC: In what years were you working on the calculations for the spin one-half versus the spin zero and the spin one, and at what point did it become clear that the actual answer was that it was a spin one-half as you had thought was possible, but nobody else had entertained that idea yet?

RFC: The reason why it was generally assumed that the meson had integral spin, that is, either zero or one, was that it was supposed to be exchanged between nucleons, and this could not be done with a spin one-half particle because there would have to be something else involved. But the idea of spin one-half particles of course was very common because the electron was known to be spin one-half. It was just that the presumption about the meson was that it was probably integral spin. My work showed that it was possible that it was spin one-half, and that was done in 1940, published in '41. But it was not until probably 1946 or '47 when a group with — I think it was Powell — in England, studying the mesons in cosmic rays, was able to find that the meson decayed and had a lifetime of some two microseconds actually. Later, it was then realized that the solution to this puzzle was that there were in fact two mesons, which are now called the mu meson (which is the one that I used to call the mesotron in the earlier days), and the pi meson (which is the one that is in fact involved with the interactions in nuclei). The pi meson indeed has an integral spin, but the mu meson it turns out has spin one-half. The pi meson disintegrates into a mu meson and a neutrino, so that there is a complicated cascade of events that did not get fully understood until the late 1940's.

IC: Could you talk a little bit more about the other part of your thesis work?

RFC: Well, one part was to calculate the electromagnetic interactions between mesotrons and the Coulomb field of nuclei, in particular what was called bremsstrahlung. When the charged particle is deflected by a nucleus, the acceleration involved in that deflection causes it to emit gamma rays, and that's called bremsstrahlung. But the other half of the thesis was to compare those calculations with what is observed in cosmic rays.

Some experimenters had observed what were called cosmic ray bursts. That is, an ionization chamber would be surrounded by lead, and it was observed that occasionally the ionization chamber would show a large burst of ionization meaning that a large number of charged particles had passed through it. This was understood as being due to a charged particle such as the mesotron interacting with a lead nucleus, being deflected, and emitting a gamma ray via the process of bremsstrahlung. That gamma ray would then multiply by making an electron-positron pair, those particles further being deflected and emitting gamma rays. In this cascade the energy goes into many, many more particles of lower and lower energies, and this is the cause of what is called a cosmic ray burst or the cascade phenomenon. These bursts had been measured, and they do give a measure of the interactions that are occurring between the mesotron and the lead in the shield around the ionization chamber. What I did was to calculate how many interactions there should be under different assumptions about the mesotron, and compare these calculated burst rates with the observed burst rate to see whether they could distinguish between these assumptions. That was the other half of my thesis project. The comparison with experiment and the close interaction of theory with experiment was a characteristic of most of the theoretical work done under Oppenheimer at Berkeley. And this was a very important part of our training, to learn the connections between theory and experiment. This kind of followed me throughout life, in that I was interested in the interaction of theory and experiment. In some theoretical schools essentially pure theory was pursued without much interest in how it connected with experiment.

IJC: How many interactions did you have with Oppenheimer? That is, how did he influence you? You mentioned these lively discussions Monday evenings, you mentioned theoretical seminars, you mentioned that he probably gave you a choice of thesis subjects, you mentioned this challenging course at the beginning. Is there anything else?

RFC: Well, we also had social interactions with Oppenheimer at his house, and these also were very important. They consisted of cocktail parties, very rarely dinner — I don't think I recall going to dinner there until he was married — but it would be cocktail parties and various social interactions. He had friends who were involved in left wing causes. This was the time of the Spanish Civil War and he knew people both at the university and outside the university, and sometimes these would be there, and so politically it was a very broadening experience to be involved with his friends. Many of his students were invited to these affairs, and the whole combination of Oppenheimer's influence, both through his teaching and research and through these social interactions, was such that it made all of his students have a very particular attachment to him, in which you might say they considered him to be very special, as if on a pedestal. He had a very special place in their lives as a result of a combination of all these things, but primarily of his own influence as a person, his own character.

IJC: Did you see him expressing very strong left-wing views at the time, or was he simply trying to help some good causes as he saw it? And was his brother there (who did join the Communist party)?

RFC: No, mostly I was exposed to a variety of left-wing causes at this time, but Oppenheimer did not himself express strong views. I did not know his brother Frank at that time, only later. Frank was not in Berkeley. So I was aware that there was a strong interest in left-wing causes, but that was all.

IJC: Coming back to Oppenheimer's classroom teaching, was that one year? That is, if you got to Berkeley in 1937 and you left in 1941, was there more than one year of course-work?

RFC: At that time he was teaching two successive courses, and so I took a course from him starting my first year there in 1937, and then I took a subsequent course from him starting in 1938. Prior to my being there and in fact during my first year there, since his courses were so difficult it was customary for people to take the same course twice. Sometimes they would work in teams: one person would listen and the other person would write, and then they would try to put things together afterward. But it was customary for people to come and take the same course twice in order to fully digest it. After my first year, however, he put an end to that, and said he would not allow that anymore, because what tended to happen was that since some of the students had already had the material before, it caused him to automatically upgrade the level (because he wouldn't want to make it too easy for them). It caused an upgrading of the course that was not desirable. So he extended the course to two years and people were not allowed to repeat. I took other courses too, and finally I had to take my qualifying examinations, which means that you study for quite a number of different oral examinations, probably three or four covering the various fields of physics. These are fairly important examinations, and each one involves studying probably for several weeks. So that took quite a period. Probably a semester would be devoted to passing the qualifying exams. That [passing the qualifying exams] permits you essentially to be a full-fledged thesis student, and when I was a thesis student I didn't take very many courses, but I took occasional courses that seemed of interest. I had actually completed my thesis work basically by 1940, and I could have graduated then. But Oppenheimer did not encourage it because he didn't have any jobs available, and if you graduated you couldn't be a teaching assistant so I would be without a job. So I was encouraged to continue for another year to write up my thesis and get it published, and by that time a job showed up and I could graduate and leave and get a job.

IJC: Oppenheimer was teaching at Caltech. How frequent was his teaching, and what was your first introduction to Caltech?

RFC: Ever since he took a job in the U.S. in the early thirties or late twenties (I forget which), Oppenheimer had a joint appointment at Caltech and Berkeley. Some years he would spend most of his time at Caltech and a smaller amount of time at Berkeley. But by the time I came along in 1937, he had developed a pattern. Berkeley was on the semester system. It started in mid-August and the first semester was before Christmas, and the second semester was after Christmas and ended, I think, in late April. Oppenheimer would teach two semesters in Berkeley. In the meantime, Caltech was on the quarter system, and by coming down to Caltech in mid-April, approximately, he could teach the third quarter at Caltech which ended in June. So he would be teaching two semesters at Berkeley and the third quarter at Caltech. At that time essentially all of his students were in Berkeley, and so when he came down to teach at Caltech in the spring his students would make a pilgrimage down with him in order to continue their interaction and take his course there and to learn about what physics was going on at Caltech. So two years I did that; probably it was 1939 and 1940, but I'm not absolutely sure.

IJC: Oppenheimer stayed in Bob Bacher's guest house in those days? And how did the graduate students find lodging?

RFC: It wasn't Bob Bacher's house at that time, it was the Tolmans' house [Richard and Ruth Tolman], and Oppenheimer stayed in their guest house. We would find rooms. I remember having a room on Blanche Street at one time, which I shared (I think) with Eldred Nelson — I'm not sure. The second time I came down, it was after Oppenheimer's baby Peter was born, so he did not choose to drive down. But he came down I guess by train with his wife Kitty and the baby, and asked Eldred and me to drive down his Cadillac convertible, which was quite an experience for us. At that time, that summer, we stayed in the Tolman guest house and I think he and Kitty were living in the Tolmans' house, and I guess the Tolmans were away. So that was the living arrangement that year.

IJC: What did you write up about your thesis, either in a published form or in an unpublished form?

RFC: I wrote up my thesis, and my thesis basically consisted of an introduction and two preprints from the papers that I had put out. So it saved me time on writing my thesis because I had already written the papers.

IJC: And what were these papers?

RFC: The ones that I mentioned. One was "Burst Production by Mesotrons," and the other was "The Interaction of Gamma Rays with Mesotrons."

IJC: Who were the co-authors?

RFC: I don't remember. Kusaka probably was a co-author on the theoretical one on the calculation, and I think the "Burst Production by Mesotrons" I did by myself without any co-author [in fact, Kusaka was the second author for both papers]. I also had a publication with Joe Keller, who was the person whom I had assisted earlier. I was a co-author on the publication which was his thesis, on the calculation of the fine structure constant.

IJC: Was there any reason at this time that you ended up in theoretical physics rather than experimental physics?

RFC: The reason was that I had not had any proper grounding in experimental physics before I came to Berkeley, whereas I had a fairly good grounding in theory, so that was the natural thing to do. Under different circumstances, a different environment, I might have become an experimental physicist of sorts. But the environment that I was in did not encourage that.

Robert (second from left) among his friends and colleagues

Chapter 4

The First Marriage, to Dagmar von Lieven

Dagmar Elizabeth von Lieven was born in 1915, during the First World War. Her parents were from the eastern part of Prussia, from a branch of the aristocratic Lieven family — one of the oldest and noblest families of Baltic Germany. This family has distinguished roots back to the 12th century, with prominent members in several countries including Sweden and Russia as well as Germany. Dagmar's father had studied engineering in Hanover. Little is known about her mother except that she spoke many languages, was involved in literary studies, and was an accomplished piano player.

The home of Dagmar's parents in East Prussia had been lost when Russia took control of the area briefly in 1914, at the beginning of World War I. Most of the German troops were fighting on the Western front, so Russia was successful in this invasion. Later, in 1914–15,

Dagmar von Lieven in the mid-1930's, at her graduation from the University of British Columbia

the Russians were forced back out and the German army advanced into western Russia. East Prussia became part of Germany again, and remained so until the end of World War II in 1945. During the 1914 Russian takeover, Dagmar's parents were forced into exile deep in Russia (presumably because they were German aristocrats), being sent to a city called Ufa just west of the southern Ural Mountains. I have been told that Dagmar's parents succeeded in taking some of their sterling silver along with them when they went into exile.

The Urals are a range of mountains that runs roughly north-south in the middle of Russia and are considered the boundary between Europe and Asia.

Ufa was (and still is) a major city and an administrative, trading, manufacturing, and cultural center in Russia. Dagmar was born in 1915 while her parents were in exile in Ufa.

A few years later, the family (now including Dagmar) fled the Communist takeover of Russia, ending up in Harbin in northeast China. This city is just south of Siberia, not very far from the Pacific Ocean. Harbin is a city of extreme temperatures and has been called "the Ice City." It was founded in 1898 with the start of the construction of the Chinese Eastern Railway, and had become a cosmopolitan city with many different languages spoken even prior to the arrival of Russian émigrés fleeing the Communist Revolution. Dagmar's father died of starvation in Harbin. Just before his death, he finished a manuscript of poetry written in archaic German handwriting, a translation of a story that had been written in Russian. He dedicated this work to his little daughter Dagmar.

When Dagmar was 13, more than a decade after the family had been forced from their estate, her mother Lydia von Lieven was able to emigrate to Vancouver, Canada. Dagmar's mother worked as a piano teacher to support herself and her daughter there, and even managed to buy a house.

Dagmar spoke several languages including German, Russian, and Italian. As a child her parents used to communicate in French in their home when they didn't want her to understand what they were saying. The language of her family was German but the language of her early environment had been Russian. She was an outstanding student, and after four years in Canada had become fluent in English. She had the second highest score in the province-wide exams in British Columbia taken at the end of high school. Her future husband, Robert Christy, won first place that year — that is how they first met. She then enrolled in the University of British Columbia (UBC) in Vancouver, specializing in French.

Dagmar and Robert kept in contact through all of Robert's undergraduate and graduate years, while they were both at UBC and while he was far away as a graduate student at Berkeley in California. When he was at Berkeley they stayed in contact mostly through letters, and saw one another only on the rare occasions when he returned to visit Vancouver. As soon as Robert graduated with a Ph.D. from Berkeley and was ready to earn a living, he and Dagmar got married, as they had been intending to do for years. This was in May of 1941. They married in a small civil ceremony as they and their families could not afford an elaborate ceremony. The young couple celebrated by going to Bowen Island for a week or two. They then went to Berkeley, where they stayed for a few months. It was there that Robert bought his first car, a 1929 Chevy Roadster, for which he paid about $50. He owned this first car for a grand total of one month — it was sold before they left Berkeley. At the end of the summer they drove in Joe Keller's car across the country to Chicago, where Robert started in his teaching job at Illinois Tech. (Joe Keller had also been a graduate student of Oppenheimer and had married just before Robert did. He owned a car that he wanted to have brought to St. Louis, where he had just landed a position at Washington University.)

Two early photos of Dagmar

Robert's Parenting of His Sons Ted and Peter

Robert and Dagmar's first child, Thomas Edward ("Ted") Christy, was born on August 30, 1944 while they were living at Los Alamos and Robert was working on the Manhattan Project (see Chapter 6). Robert had not experienced parenting from his father or his grandfathers, as they had all died before he was old enough to remember them. It was very important to him to become a good father to his own sons. Robert found that he greatly enjoyed being a father and devoted as much time to Ted as he possibly could. Sometimes he would go home to his wife and son during his lunch break. After work he would come home as soon as he could, and would join his family for dinner and play and read with Ted until Ted's bedtime. After that he would work on his physics.

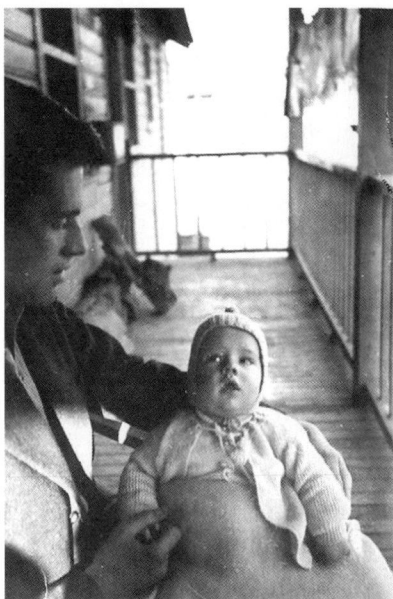

Robert with his first son Ted, in the winter of 1944–1945

Robert and his second son Peter playing outdoors during a snowy winter in the late 1940's

Robert and Dagmar's second child, Peter Robert Christy, was born in Chicago on May 4, 1946. This was nine months after the successful test of the "Christy Gadget" — the first atomic bomb — at Alamogordo (see Chapter 6). Robert was as happily and deeply involved with Peter as with Ted.

It was always very important to Robert to teach a love of the outdoors to his sons and to provide them with a good education. He taught them how to swim and play baseball, took them hiking and skiing, and took them on various camping trips and road trips. When he had consulting work in La Jolla, California, he took his whole family there to enjoy the ocean. He taught his sons to swim in the ocean, and to body-surf.

Robert designed and built a swimming pool for them almost singlehandedly at their home in Pasadena (at 2810 Estado St.). Unfortunately, this pool and house were demolished when the City of Pasadena built the 210 freeway there a few years later and the family had to move out.

Robert with Ted and Peter

When Robert was on sabbatical at Princeton, he introduced his sons to the theatre and shows of New York City. Years later, he organized a two-week road trip with his sons through the Rocky Mountains of the western US and Canada. They had great fun hiking and camping.

Robert built a car in their garage with his younger son Peter, and introduced him to the basic construction of computers. Peter was interested in computers from an early age. In Robert's 2006 interview with Sara Lippincott, he said, "I had helped him put together the essence of a computer with some relays that I got. And those relays would operate, and they would do 'yes/no' commands. He got into computers very early."

The swimming pool that Robert built, while it was still under construction

Robert's Memories of His Sons

In February 2011, Robert recounted some of his memories of his sons Ted and Peter:

> All the time I had, I played with them [Ted and Peter] and enjoyed it very much. I would have them on my long knees.
> Later as they got bigger I went hiking to Yosemite. I liked doing those things and I took them along. We would have rooms in cottages and have cooked meals. We would spend the day hiking around. This was when they were little and up through the time when they were teenagers. I did these things with them because I liked to do them myself. I enjoyed the outdoors.

Robert with Peter

Then I came to Pasadena and I was busy until 10 or 11 at night. I would play with them at night and then go back to work. On the weekends, I went outdoors as much as I could and I took them along. Their mother was not an outdoor person so she would not come along.

Growing up, we all came to La Jolla for a couple of visits (all four of us) lasting a couple of weeks. There were meetings held at La Jolla, and we stayed in a motel and went to the La Jolla shores for swimming. The family would be at the beach and I would come in my time off. Their mother stayed on the beach and I went into the waves with the boys. The purpose was to enjoy the waves. We had boogie boards. We did very elementary rides of the waves. I never learnt to stand on a board. Also Tommy Lauritsen and his wife and kids would be there.

Occasionally, I helped with school work in math or science. They were very good students. I often had lectures and could not be there in the mornings. I did not eat in the Athenaeum at that time and I would either come home or go to Chandler Dining Hall. As soon as I could, I would go home to do my homework.

I did not have an office in our home, but I had some space that I used as an office. I would be playing with the children and help them do their homework and put them to bed, and then do my own work.

I built a swimming pool for them but it turned out by the time the pool was finished, they were 16 and they could drive and all they wanted was to drive off to the beach with girls. (They did not tell me that, but that is what they did.) I cleaned the pool every week. While the pool was not a major attraction, it was used for general recreational swimming. I swam in it, and the boys too. I taught myself how to dive in it.

The boys went to public schools because they were very good schools at that time. The boys were very good students. I encouraged

Robert with Ted

Robert with his sons

them. I did not specifically encourage them to do physics; I showed them that I enjoyed science, but I did not tell them what to major in. They were much better writers than I was. My general advice was to do the things you enjoy and like to do. Ted started off in physics, probably to emulate me, but switched to chemistry and finally went into biology. I did not force him to go into chemistry or biology; I really encouraged him in doing what he wanted to do.

Peter did not go into physics, he went into computers. I helped him in his interest. We built a very simple computer made out of relays that simply said "yes" or "no." Peter and a friend of his programmed a stunt for football games where they trained members of the audience to hold up cards that created words. I also helped him obtain a job at Caltech.

When Ted was 16, I went to Germany. I bought a Volkswagen for the family and I spent a month or more driving around, and I did the whole trip on five dollars a day. I was looking forward going back home. I stopped in France too and saw those cave paintings of thousands of years ago. The car was shipped over from Brussels, while I flew. Of course I was happy to be home.

When I was in Princeton, I went out of my way to show the boys the arts and highlights of New York. There were some problems in changing the schools from Pasadena to Princeton.

It was important for my children to pursue higher education as much as they were able and we would pay their bill. It is bad to start life after college with bills. In the Depression years when I was a college student I had no money. When we were in Princeton in 1960, I told the boys if they wanted to pursue higher education we would pay for it.

Haverford College nurtured students: it was a liberal arts college. Ted had no problem getting admitted because he had good grades. He made many contacts and got to know

At the La Jolla beach

Robert with Ted

everyone. We paid for it. I don't recall how. I was a full professor at that time. It was not easy.

Peter decided on Harvard. He did not get in initially, but due to his computer stunt he got in the second time he applied. Peter had made a computer out of relays, and Harvard was so surprised that he was admitted.

As a family we had three Volkswagens. As their mother did not drive there were three Bugs for me and the boys. The one I shipped from Brussels was mine. I did not encourage motorcycle driving. One time we went to Philadelphia, to see the Liberty Bell.

After Princeton, we came back to Pasadena. A year later, in 1962, Ted was able to drive one of the Volkswagens. We [Robert and his sons] visited a lot of National Parks — Zion, Bryce, and Yellowstone National Park. Then we went up to the National Park in Montana. There is a Glacier National Park there. We drove up to Canada through the Rocky Mountains, then to Vancouver — and driving in BC, there is the most beautiful scenery everywhere. We never stayed in motels, we were in camp grounds with sleeping bags. All three of us enjoyed that trip very much. One of my rules was I liked to be outdoors but I did not like to cook, so we slept outside without tents and we went to eat in tent cabins where there was a central cookhouse. It was a very common arrangement.

Life was quiet after Ted left. Peter was neither easy nor difficult. The boys were occasionally rebellious, but not all the time. I am sure they smoked marijuana but I did not know — they hinted at it later. They had friends, they did not avoid girls (but that was private), and they always came home for dinner. I allowed Peter to use my credit card to buy books and he bought a ton of them and I am sure he read them all.

Ted injured his back in high school while he was training for football. That was the first one. He got out of his problem exercising and resting. Peter was kind of roundish when he

Robert with Peter

was born, and he always kept some extra weight. They always liked to cook and they taught themselves. Their mother liked to cook too. I tried to teach tennis, but they had no real interest in it. I taught them to swim. I went to local mountains and rented equipment and taught them how to ski. They did not play any musical instruments, and neither did I. I was beginning to take piano lessons when I was ten, when my mother died. I would have liked to play it better. I had a piano at home, and I liked to keep one because it encourages people to try.

The first time I went to ski with Ted, we skied off the road and Ted hurt himself in the knees. That was the ending of skiing for a while. Peter came along too. We learnt very slowly. Peter and I skied at Mammoth many times. Decades later, Ted joined us skiing at Mammoth with his wife and child.

We did not celebrate birthdays in a big way. We celebrated Christmas, but not Easter. I cooked Grandma's turkey for Thanksgiving.

Ted dropped out of Berkeley when he injured himself. Then he went in the mountains and became a hippy. So I suggested him to become a medical doctor so he could help people. He went through medical school at USC [the University of Southern California], and became a general practitioner.

Ted and Peter in College and as Adults

It was very important to Robert that his sons come out of college without the burden of college debts that so many students incur. He also wanted them to live better than he had been able to while at college. Despite the family having only a single income, Robert accomplished this goal. He managed to pay for their tuition and living expenses, and even managed to buy each of his sons a Volkswagen.

Peter was admitted to Harvard as an undergraduate. The admissions committee was particularly impressed by his computer project. He studied a combination of engineering, mathematics, and computer science. He went on to Berkeley as a graduate student in electrical engineering. After several years at Berkeley, he left academia and went to work for the Digital Equipment Corporation. He moved from one company to another, including IBM and Apple. He and a few other people founded their own computer company, dealing with parallel computing. Until 2012, he and a partner ran their own successful consulting company. Recently he has taken up a position as a technical consultant with an international company.

Robert said about his older son Ted:

Ted tried just about everything. He started off [at Haverford College, a liberal arts college] in chemistry because he didn't want to clash with me on physics. But when he was there he went into physics and tried that, and then into biology and tried biology. And I think he enjoyed biology more. He left Haverford and went to Berkeley as a graduate student in biology in the late 60's. He dropped out — which was more or less typical for the period — for a couple of years, until I managed to persuade him to get back and go to medical school. And he went to medical school at USC [the University of Southern California] then and finally became a family practice physician, which I think he's enjoyed very much, and has been a very useful member of society.

In a letter written to me in November 1970, after he and his first wife had separated and were in the final stages of their divorce, Robert described the personalities of his two sons, who were in their mid-20's at that time:

> You say you would like to meet Peter and Ted when you come this spring. I would be happy to arrange it. It is more straightforward with Peter who lives just on the other side of town and comes to visit occasionally. You should understand that Peter is, however, very inquisitive and no doubt jumps easily to conclusions. He visited one day when I was in the middle of writing you a letter — so I picked up what was lying around and put it away — he asked no questions. A week or two ago he came over and I helped him with some work installing a radio in his car. Later I went across the street to exercise in the school ground and after I came back he said he had gotten a knife of mine from my desk drawer where he knows I keep such things. Your pictures are also in that drawer only a little deeper in. He did not say he had seen anything unusual and I did not ask if he had seen pictures of a beautiful woman I keep there. But with Peter I always assume he knows anything there is to know. So when you meet him you should anticipate that he will jump to conclusions. Many years ago he hinted to me he had overheard some discussions my wife and I had had, and knew all was not well.
>
> Ted is quite different, he does not pry as Peter does but he is very sensitive to people and will fairly readily understand situations. Did I tell you last September that he was dropping out of graduate school this fall and was trying to find himself and his future? He is still living in Berkeley however. But it is an easy airplane ride to Los Angeles so there is no problem about getting together when either of us wishes. Thus in meeting them, there will be inferences drawn (which does not disturb me but you should understand). They, however, do not spread their knowledge and nothing they know would reach Caltech.

An Unusually Amicable Divorce

Dagmar and Robert had rather different interests. Many years later Robert concluded that their years apart while he was a graduate student, during which they stayed in touch only via letters, kept them from realizing those differences that had not been obvious during their previous acquaintance at UBC.

Robert preferred active recreations such as skiing, swimming, and tennis. He would have liked to dance, and in fact had taken dancing classes at Berkeley. On the other hand Dagmar preferred quiet activities such as reading, painting, and other artistic hobbies. Dagmar preferred not to dance and rarely travelled. She did not join Robert when he went to conferences abroad, nor on his camping vacations with his sons. This resulted in neither of them having companionship in most of the activities that they enjoyed.

Dagmar and Robert separated in the late 1960's, nearly three decades after they were married. They were divorced not long after that; it was finalized in early 1971. It was Dagmar who had suggested the separation and the divorce, although Robert had also been painfully aware for many years that all was not well in their marriage.

Their divorce was highly unusual in that they treated each other so fairly, each one wanting to ensure that the other would be able to continue living reasonably thereafter. Their assets consisted of their three-bedroom debt-free house and Robert's life savings, which were comparable to the net value of the home. He asked his wife Dagmar what she wanted, and she said "the cash." So he gave the entirety of his life savings to her. In addition he volunteered to give her a large fraction of his professor's salary from Caltech so that she could also live reasonably. When they separated in the late 1960's her after-tax income would have been roughly twice his, leaving him barely enough to live on. It was only after he unexpectedly became Provost (Academic Vice-President) of Caltech in February 1970 that his salary reached a level that he could live on comfortably, at the same level as before their separation. In addition he offered her anything she wished from the contents of the house. She took only half of the furnishings, insisting that he keep enough so that he could also live reasonably. Instead of retaining separate lawyers for the divorce proceedings as is normally done, Robert and Dagmar felt that they could work things out between them. They did not wish lawyers to cause hard feelings, so they shared the same lawyer for the necessary legal procedures.

Just over a decade later, when Dagmar died in early 1982, it was revealed that she had appointed Robert to be the Executor of her will and had granted him any of her furnishings that he might wish. Her other assets went to their sons Ted and Peter.

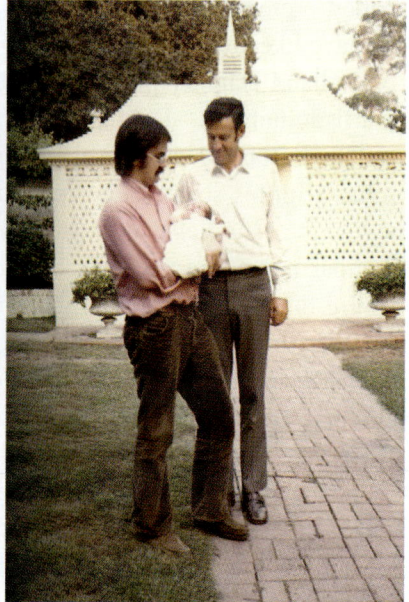

Above left: Ted, Robert, and Peter at Robert's wedding to Juliana in 1973. Ted is almost 29 years old, Robert is 57, and Peter is 27. Above right: Peter and Robert with Robert's first daughter Ilia in 1974: ages 28 years, 58 years, and 3 weeks, respectively

Chapter 5

The Very First Nuclear Reactor

An Unfulfilling Job Teaching at Illinois Tech in 1941

After the exciting years of physics as a graduate student with Oppenheimer at Berkeley, coming to Illinois Tech in the summer of 1941 after obtaining his Ph.D. was a let-down for Robert. However, it was the only job available to him that year. As Robert put it,

> Essentially, jobs were obtained by the system that Oppenheimer would receive notice that someone had a job available, and Oppenheimer would figure out what might be possible. The one thing that he knew about was the job at Illinois Tech (which I took), probably because someone had written him asking him if he had any students who would be available. There conceivably were other jobs available in the country, but they were not known to the group in Berkeley.

Joe Keller had taken a job at the Washington University in St. Louis the year before. George Volkoff had gone back home to Vancouver to take a position at the University of British Columbia (UBC). Robert was glad that his good friend had found a position there, though he himself had no wish to return to Vancouver. Eldred Nelson accepted a job at Berkeley, not in theoretical physics, but using his knowledge of theoretical physics to assist Lawrence in the experimental lab. Oppenheimer felt that Robert's Illinois Tech was a better job because it was purely in theoretical physics.

Robert was to teach three classes, each one nine hours a week. He had a very heavy teaching load with 27 hours of teaching a week for 11 months of the year — typically much more time is spent preparing lectures and creating and marking assignments than in the lectures themselves. Additionally there were three or four graduate students for Robert to supervise and to provide thesis work for. His salary would be $200 per month, but payable for 12 months of the year, unlike his graduate fellowship. One month a year he was free to work on his own research.

Robert's normal work day consisted of working at least until 11 p.m. His young wife was left largely to herself as there was almost no time or energy left for a life outside of physics. This was fairly typical for most young physicists.

Illinois Tech at the south end of Chicago was surrounded by a rough district. In Robert's words:

> Illinois Tech was on the south side in a rather difficult district. It was right next to the railroad tracks, and there were steam locomotives all the time so that there was coal dust being blown into your windows all the time. You had to keep your windows closed. The "El" (which is the elevated railroad) ran nearby and it was considered dangerous to be there at night. So we didn't work at night at Illinois Tech because someone not too long before had lost his overcoat by being held up at knife-point at night.

Thus Robert and Dagmar decided to find a place to live in a safer area further north towards the University of Chicago.

It was of key importance to Robert to keep abreast of what was going on in physics by going to the physics seminars at the University of Chicago. He travelled there by bus as he did not have a car and bicycle travel was too dangerous. He had been accustomed to the dynamic group at Berkeley and was used to asking pointed questions whenever relevant. He was soon noticed by the physicists at the University of Chicago because of these questions. The Manhattan Project had just been started, and they were looking to hire a new theoretical physicist. They took note that Robert was a recent graduate from Oppenheimer's group at Berkeley, and saw that he was brilliant and well-informed in physics.

Background: The Discovery of Nuclear Fission in Germany

The initial motivation for the U.S. atomic bomb project came from the fact that U.S. scientists had learned that their colleagues in Germany, who had recently come under Nazi control, were working on military applications of nuclear fission.

Nuclear fission had been discovered in Germany in late 1938 to early 1939. In 1938 in Berlin, Otto Hahn and Fritz Strassman discovered that barium sometimes resulted from a neutron colliding with uranium. In a letter to Lise Meitner, Hahn described this as a "bursting" of the uranium nucleus. Meitner, a physicist of Austrian Jewish origins, had already left Nazi Germany and was living in Sweden. She received this letter during the 1938 Christmas holiday when her physicist nephew Otto Robert Frisch was visiting from Copenhagen. In early 1939 Meitner and Frisch wrote a paper theorizing that the uranium nucleus had split in two, and estimated the amount of energy released — a huge amount, compared to the energy released in chemical reactions. It was Frisch who coined the term "fission" to describe this new process.

Origin of the German Nuclear Project

Shortly thereafter, in early 1939, informal German nuclear energy projects began. Heisenberg was one of the principal scientists leading the research and development. Three scientists at the University of Göttingen — Georg Joos, Wilhelm Hanle, and Reinhold Mannfopff — became known as the "Uranium Club." Hanle presented a paper in a colloquium that April on the use of uranium fission in a *Uranmaschine* ("Uranium machine," i.e., a nuclear reactor). This first "Uranium Club" was discontinued a few months later in August 1939 when the three scientists were summoned for military training.

It turned out that as early as April 1939 German scientists had notified Hitler's Ministry of War and Ministry of Education that nuclear chain reactions had potential military applications. The second "Uranium Club" was initiated on September 1, 1939 (the day the war began) as a formal nuclear energy project controlled by the military.

Origin of the U.S. Nuclear Project

Research into nuclear fission was also starting up in the U.S. Leo Szilard had been able to convince Columbia University to give him some laboratory space but was short of funds to do research. He had been considering the possibility of a nuclear chain reaction in light elements (which of course turned out to be impossible) until he heard of the discovery of uranium fission in 1938. As described in a manuscript by Norman F. Boas, Szilard then approached his friend Benjamin Liebowitz, a businessman and physicist, for a $2000 loan to enable him to start research into nuclear chain reactions. Liebowitz promptly loaned him this money. These were the first American funds actually spent on the chain-reaction concept. (The author Norman F. Boas was the son of Ernst P. Boas, M.D., who was a friend and associate of Benjamin Liebowitz.)

The famous Einstein-Szilard letter was sent to President Roosevelt on August 2, 1939. Although the letter itself had only Einstein's signature, it had been largely composed by Leo Szilard with some help from Edward Teller and Eugene Wigner. Due to Roosevelt's preoccupation with Hitler's invasion of Poland he did not receive the letter until August 11, 1939. The letter warned that it might be possible to produce nuclear power and perhaps even an extremely powerful atomic bomb. It pointed out that good sources of uranium ore existed only in Canada, the former Czechoslovakia, and the Belgian Congo. It also noted that Hitler had stopped the sales from the Czechoslovakian mines that he now controlled.   The letter recommended that the U.S. government should get involved in the U.S. atomic research effort.

Oppenheimer, Teller, Fermi, Wigner, Weisskopf, and many other prominent nuclear physicists in the U.S. had previously worked in Göttingen. There they had become acquainted with a number of the outstanding nuclear physicists who were still working in Nazi Germany, such as Heisenberg, von Weizsäcker, and Hahn.   The U.S. physicists thus were well aware of the brilliance of these German scientists, and were worried that they might be capable of producing a powerful nuclear bomb for use in the war, which would be a disaster for the Allies. During the first third of the 20th century, Germany had been recognized as being preeminent in physics. The American and British scientists feared that a dictatorship like that of the Nazis might create an organized, determined, and eventually successful atomic weapon project.

The Early British Nuclear Project

The precursor to the U.S. Manhattan Project started in late 1939 as a small research program. At that time it was assumed that any nuclear bomb would require many tons of uranium and thus would be rather impractical. However, in 1940 (Otto) Robert Frisch and Rudolf Peierls, who had been working in the British nuclear research program since 1939, wrote the Frisch-Peierls Memorandum. In this document they described how extracting the fissionable isotope uranium-235 from natural uranium could produce an immensely powerful detonation using a very much smaller amount of uranium. They estimated that only a few pounds of uranium-235 would be needed, as opposed to several tons of natural uranium — in other words, an atomic bomb might be feasible. As described in a 1998 letter from A. Murphy (a British historian) to Professor Christy, in June or July of 1940 a meeting was set up between Peierls and Lindemann — Lindemann had Churchill's ear on scientific matters. At that meeting, Peierls described to Lindemann how a practical atomic bomb was now possible.

The Frisch-Peierls memorandum reached the U.S. in 1941. It was a key trigger for the establishment of the Manhattan Project, whose goal was to produce an atomic bomb.

The Manhattan Project Begins

President Roosevelt approved the atomic project on October 9, 1941, with the Army to hold principal responsibility. (This was two years after the Nazis had started their atomic project, also under army control.) By late 1941 work was going on at four centers across the U.S.: Columbia University in New York City, Princeton University in New Jersey, the University of Chicago, and the University of California at Berkeley. In June 1942 Colonel James C. Marshall was appointed head of the Army's part of the Project, initially titled "Development of Substitute Materials." Colonel Leslie R. Groves thought that this name was too descriptive. The new name "Manhattan Project" was chosen because Groves was deputy head of the Construction Division of the Army's Manhattan Engineering Corps.

Marshall was not thought to be pushing the Project ahead strongly enough. On September 17, 1942 Groves was promoted to Brigadier General and replaced Marshall as head of the Manhattan Project.

A scientist would be needed as director of the secret laboratory that would actually build the atomic bomb. The eminent scientists Urey, Lawrence, and Compton could not be spared from their positions as heads of their respective laboratories. According to Robert Christy, Carl Anderson of Caltech was also considered, but was thought to be unable to organize an interdisciplinary project of this magnitude that would involve not only physics but also chemistry, metallurgy, ordnance, and engineering. Compton suggested J. Robert Oppenheimer. In October 1942 Groves became convinced that Oppenheimer would be suitable as director, in spite of not having a Nobel Prize (unlike the other four) and in spite of security concerns due to his connections with known Communists. Groves decided that Oppenheimer had an "overweening ambition" that would enable him to push the project to a successful conclusion, an ambition that he felt was lacking in the other scientists.

Incidentally, although Oppenheimer immediately started work on setting up the secret laboratory, choosing the Los Alamos site (near his own ranch) in November 1942, he did not actually receive his security clearance until July of 1943 — the security requirements were personally waived by General Groves. In choosing to take over the Los Alamos Ranch School, Oppenheimer noted as one consideration the "natural beauty and views of the Sangre de Cristo Mountains," which he hoped would inspire those who worked on the Project.

Robert Joins the Project

Wigner was the head of the theoretical group of the Manhattan Project at Chicago. He hired Robert in the late fall of 1941 to start to work with him in January 1942 in a full-time research position. He offered Robert the same pay that he had had in the far less desirable teaching job at Illinois Tech. It is amusing that it cost twice as much to hire someone to replace Robert at Illinois Tech since it was difficult to find a replacement in the middle of the teaching year — Wigner actually ended up paying Illinois Tech for this extra administrative expense. As Robert described it:

> The Project was just getting organized then, actually. Prior to the fall of 1941 the Project had consisted of separate projects. There was a project in Columbia, where Fermi was working [Fermi had immigrated to the U.S. in 1938, from Rome], and also Harold Urey was working on isotope separation; Fermi was working on chain reactions. There was a project in Princeton where Wigner was working on the theory of the chain reaction, and a group under Bob Wilson was working on a new technique he had invented for isotope separation. There was some work going on at the University of Wisconsin and the University of Minnesota (some neutron experiments), and there was some work going on at Chicago. But none of this was brought together into a real coherent scheme until the fall of 1941 when it was decided to put the whole thing under the control of the Manhattan Engineering District of the United States Army, with then (I think) Colonel Leslie Groves as the boss who would be running the whole organization.
>
> At that time it was decided to kind of concentrate things, namely, the isotope separation work would continue under Urey at Columbia University; the chain reaction work would be concentrated and combined at the University of Chicago under Arthur H. Compton, and Fermi and Wigner would move to Chicago to work on the chain reaction project; and the fast neutron project was still being coordinated by John Manley and Robert Oppenheimer, but was continuing to go on at various universities like Wisconsin, Minnesota, and Berkeley.
>
> When I went to the University of Chicago they had just brought the work together there, to start what was called the Metallurgical Lab (or "MetLab") as

that branch of the Manhattan Project; and in fact Fermi had not arrived there at that time. I worked initially with Wigner on the theory of the chain reaction, and then some months later when Fermi arrived I worked with Fermi on some of the experiments he was making on small chain-reacting systems which were too small to fully chain-react (called exponential piles).

(Fermi and his team moved from Columbia University to the University of Chicago in March 1942.)

Robert was with Wigner in the theoretical group for only a few months. They were working on the theory of the uranium chain reaction with a graphite moderator, which Robert learned about from Wigner. Robert recalled:

I probably knew what the general purpose of the Project was, but I don't have a clear recollection of when I was told all of these details. When I was first hired they didn't know I was an alien (I was still a Canadian), and they then had to hurry up and clear me so I could be given the secrets I had already been given when they didn't know I was an alien.

I gather I knew what was involved, although I don't explicitly remember exactly when I learned that or anything. What was involved was to make a uranium bomb, and there were two approaches to it that were being pursued. One was to separate isotopes: uranium-235 could be used for that, it was found. The other approach was to use plutonium-239. Neither of these existed as pure substances. One part of the project was to separate uranium-235, and that was under Urey, and eventually huge isotope separation plants were built at Oak Ridge, Tennessee. The other part of the project was to produce plutonium by a chain reaction which would involve ordinary uranium, and as we found out it could be done with a graphite moderator. The uranium-235 would fission and keep the chain reaction going, and some of the neutrons would be captured by uranium-238, which [usually] would not fission. But the uranium-238 plus a neutron gives neptunium-239, which decays by beta-decay to plutonium-239. And so plutonium is created in the chain-reaction as a parasitic process accompanying the chain-reaction. Eventually enough plutonium accumulates in the uranium that the uranium can be removed from the reactor and chemically processed to remove the plutonium, since it is a different chemical element. Plutonium can then be concentrated and manufactured into a metal — and that's what was eventually shipped to Los Alamos for construction of the plutonium implosion bomb.

Fermi approached Wigner and asked if Robert could work full-time with him in the experimental group. Wigner agreed, and Robert started work with Fermi in early 1942. Robert recalled this period:

I switched to working with Fermi on both some calculations and some experiments that he was doing on small chain-reacting systems which were too small to fully chain-react — but they were called exponential piles. They were essentially the prelude. He kept on building new exponential piles with new materials, purer graphite and purer uranium, and adjusting lattice separations, until he could prove that he had a system which, if extended to a larger size, would become fully chain-reacting. I was working with him in the later stages of that.

When I arrived, the other theoretical person working with Wigner at the time was Alvin Weinberg, whom I have known ever since. I still meet with him from time to time on various committees. He had been a student in theoretical biophysics at Chicago and got impressed into the [Manhattan] Project there. There was Sam Allison, a more senior person probably ten years or so older than I was, who was a faculty member at the University of Chicago. There was Professor Compton and there were a number of young people. Ed Creutz was, I think, an engineer, but he was responsible for some of the technical aspects — he was involved in how to prepare uranium for compressing it into lumps for the reactors. There were people involved in the chemistry there that I ran into but did not know very well. I think Glenn Seaborg was there, and as I say there were a variety of people that I haven't interacted a lot with since but whom I did run into there. In the theoretical work there were not very many. I was one, Wigner, Alvin Weinberg, and Fermi (both theory and experiment of course).

My work was to do whatever I was asked to do. I did some work on trying to calculate the effects of different lattice spacings on the reproduction constant of a graphite-uranium system. I did some calculations on the control of a chain-reactor — control mechanisms, and how they respond to control. I did various jobs, whatever was assigned. I don't think most of it was written up, not in a form that I'm aware of at least.

One exception to this lack of formal documentation was a seminar that Fermi gave there and which Robert wrote up. This eventually was published among Fermi's papers.

The Creation of the First Nuclear Reactor, Chicago Pile-1

Wigner and Fermi's goal was to demonstrate the feasibility of a self-sustaining nuclear reaction. To do this, a pile of pellets of natural uranium and blocks of graphite was assembled under the concrete stands of the University of Chicago's abandoned Stagg Football Field (this location had previously been an enclosed racket court). This was the only place at the University of Chicago large enough to hold the pile. The graphite blocks and uranium pellets were arranged to form a flattened sphere 25 feet across and 20 feet high. Wooden supports filled in the lower corners, making it appear more like a square.

The natural-uranium pellets in the pile contained mostly U-238 (99.3%), with a small amount (0.7%) of U-235. The U-235 occasionally decays by spontaneous fission, producing a number of daughter products, including a few fast neutrons. The key to a chain reaction is that if one of these neutrons is captured by another U-235 nucleus it will usually fission immediately, releasing more neutrons. On the other hand, if a neutron is absorbed by the predominant U-238 no neutrons are usually produced. A self-sustaining nuclear chain reaction is one in which, on average, each neutron creates more than one additional neutron, so that the number of neutrons increases exponentially.

The graphite blocks were important because they acted as a moderator: they slowed down the neutrons produced by the uranium. Slow neutrons are more easily absorbed, and thus are less likely to travel far enough to escape the reactor pile and be lost to the chain reaction. Cadmium rods were used as a control system. Cadmium absorbs neutrons, thus eliminating them from the chain reaction. Pulling the cadmium rods out from the pile meant less neutrons were absorbed, allowing the chain reaction to proceed; inserting the cadmium rods allowed the chain reaction to be shut down. In Robert's words:

> It had been discovered (as much as anything by Leo Szilard, I think) that an impurity — boron — in graphite was what was inhibiting the chain reaction; and they made great efforts to go to manufacturers of graphite in order to get the graphite manufactured boron-free. That pure graphite began to become available in the summer of 1942.
>
> It is noteworthy, perhaps, that the German atomic project had considered graphite much earlier for this purpose [as a moderator to slow down the neutrons], and had discarded it as being unworkable because they did not understand that it was impurities that was causing the graphite to absorb too many neutrons. They thought carbon must have too large an absorption cross section and so they discarded carbon and went to heavy water. Their problem there was that they couldn't get heavy water because it came from Norway, and the supply of heavy water was always being cut off by commando raids on the Norwegian plant. [Heavy water is water where the two hydrogen atoms are actually deuterium, the rare heavier isotope of hydrogen: D_2O rather than H_2O.]
>
> In any case, Fermi and Szilard found out that boron was indeed an absorber, and when they were able to get graphite made without boron then the graphite turned out to be good enough [as a moderator] to make a chain reaction. At the same time they also had to get pure uranium. First they had uranium oxide, and it was possible, but it was not nearly as good as uranium metal. But uranium metal was very difficult to work with, and the ways of casting uranium metal had to be developed. All of these things — making pure uranium metal and making pure graphite — were the things that stood in the way of making a chain reaction.

Sometime in the early fall they finally got materials that were good enough for a chain reaction.

I was involved with Fermi in the first exponential pile that demonstrated that the reproduction constant was greater than one, and therefore that it could be chain-reacting.

This exponential pile measured eight feet square by eleven feet high. Note that this test pile did not itself support a full chain reaction. Its crucial importance was that it demonstrated that a larger pile *would* be able to support a self-sustaining nuclear chain reaction. Robert said:

> Later I also assisted in the construction of the first chain-reacting pile [Chicago Pile-1], which was built in Chicago. Many people on the project there worked hard, three shifts a day, machining graphite in the woodshop at the University. With graphite dust all over the place it was a rather dirty occupation. Others were involved in making the uranium lumps. It was a joint project of many people working together to put together this reactor. Fermi's two "lieutenants" (you might say), Herb Anderson and Wally Zinn, were both there. I think Art Snell was there (a nuclear physicist); Ed Creutz, probably (his main responsibility was in preparing the uranium); Al Wattenberg was there; I was there. But there were many others. There must have been more than twenty who were present who had played a role in one form or another in preparing the first chain-reaction.

In fact there were more than twice that many present at the first successful test of Chicago Pile-1 on December 2, 1942. A partial list in 1958 by the Atomic Energy Commission names 42 participants, and there are 50 on a recent (2007) list by the Argonne National Laboratory. (Incidentally, Art Snell and Ed Creutz were not on these lists.) The only contemporary written record of who was actually present is the straw cover of a Chianti bottle that Wigner had brought for the celebration, which many of the other participants had signed.

Robert reported that as the pile was assembled Fermi was very careful to frequently measure the amount of neutrons produced. This grew sharply as the pile approached "criticality" (i.e., a self-sustaining nuclear chain reaction). In addition to the cadmium control rods, when this criticality was approached, Fermi had three men stand on top of the 20-foot-high pile. According to Prof. Charlie Campbell of the University of Southern California, one of these men held an axe ready to cut a rope that would drop a control rod into the pile. The other two men stood beside glass jugs of cadmium sulfate solution, which they could smash in an emergency to release the solution and damp the reaction (though this would ruin the reactor permanently). Fermi had convinced everyone that his calculations — and the above precautions — were reliable enough that there was no danger of a runaway chain reaction or of an explosion at the University of Chicago, which would have been catastrophic in one of the most densely populated regions of the U.S.

Layers 18 and 19 during the construction of the full-size atomic pile, Chicago Pile-1, which had 57 layers in total when completed; note the uranium oxide pellets inserted in holes in the graphite blocks of layer 18, while layer 19 consisted of pure graphite blocks (LANL image)

The most important factor that allowed control of the nuclear chain reaction was the fact that after a neutron is absorbed, fission and production of more neutrons does not always take place immediately. About one percent of the time the neutrons are not produced for several seconds to a minute. Thus if a nuclear chain reaction is only just achieved, i.e., the amount of uranium is less than one percent above the critical mass needed to sustain the

Sketch by Melvin A. Miller of the first nuclear reactor, Chicago Pile-1 (LANL image)

chain reaction, the chain reaction grows with a timescale of seconds to minutes rather than microseconds or milliseconds. This allows time to manipulate the control rods to dampen down the reaction again and keep it from growing out of control.

It is also noteworthy that in this very first nuclear reactor there was no radiation shielding and no cooling system. The pile was indeed radioactive, primarily from the neutrons that managed to escape. However, the amount of radioactivity was not large enough for them to invest the time and expense to build shielding. The nuclear reactions did of course produce energy which ended up as heat in the reactor — but only about half a watt. Since this reactor was a "proof of concept" it was never allowed to get hot enough to require a cooling system.

It was on December 2, 1942, about a year after the work on reactor design started, that a self-sustaining nuclear reaction was first demonstrated in this pile, later designated "Chicago Pile-1." This was the first nuclear reactor in the world. The painting "Birth of the Atomic Age" of this historic event (see the following page) was not created until fifteen years later. The painter, Gary Sheahan, sent preliminary sketches to those who were thought to have been present asking for corrections of the layout of the room and for names of those who had been present. The initial list of 42 names that had been supplied to him by the Atomic Energy Commission was updated to a total of 50 participants, though only about half of them are shown in the painting. The layout of the observers' gallery at the left was also quite different in the initial sketch that had been sent to Robert Christy, when compared to the version in the final painting.

The Chicago History Museum painting titled "Birth of the Atomic Age," by Gary Sheahan. This painting shows Chicago Pile-1 at the historic event when a self-sustaining nuclear chain reaction was first demonstrated. It was painted in 1957, fifteen years after the event; no photograph of the pile exists.

Founding of the Hanford Reactors

With Wigner, Robert worked on the design of the first Hanford reactors that were
used for plutonium production. The Hanford site was first established in 1943 on
the Columbia River in the south-central area of the state of Washington. The
Hanford B Reactor was the first full-scale plutonium production reactor in the
world. It produced the plutonium for the first atom bombs: the very first one (the
"Christy Gadget") that was tested at Alamogordo, as well as the third one (the
"Fat Man") that was dropped on Nagasaki. Later, during the Cold War, the
Hanford site was expanded to include nine nuclear reactors and five large
plutonium processing complexes. It produced the plutonium for most of the
60,000 weapons in the U.S. nuclear arsenal. (Note that the more powerful H-
bombs developed by Teller in the 1950's still require a plutonium bomb as part
of their design, to set off the hydrogen fusion.) Today Hanford is mostly
decommissioned, but remains the most contaminated nuclear site in the U.S. due
to the radioactive waste left behind by decades of plutonium production. The
nation's largest environmental clean-up project has been working on removing
the contamination.

   As described in Chapter 7, after the war was ended by the first atomic bombs,
Robert was active for the rest of his life in the efforts to stop nuclear weapons
proliferation.

Founding of the Argonne Reactors

With Fermi, Robert later worked on the first Argonne nuclear reactor. This was
the direct successor of Chicago Pile-1 (which had been situated under Stagg Field
at the University of Chicago as mentioned above). Experiments with a nuclear
reactor were considered to be too dangerous to conduct at a university in a major
city. Therefore the experiments — and the nuclear reactor — were moved in
early 1943 to a site in the Red Gate Woods about 25 miles southwest of Chicago.
The reconstructed Pile-1, now with a radiation shield, was renamed "Chicago
Pile-2."

   Robert did not work on the pile at this time. He had been chosen by
Oppenheimer to be one of the early recruits to work on the bomb project at Los
Alamos (as described in the following chapter). However, as described in
Chapter 8, Robert returned to the University of Chicago after the end of the war

in February 1946. There he spent several months working with Fermi again on nuclear reactor designs.

In July of 1946 (two months before Robert left Chicago for Caltech) the repositioned reactor laboratory in the Red Gate Woods was formally chartered as the Argonne National Laboratory — the name Argonne came from the name of the surrounding forest. The experiments performed there were intended to result in nuclear power reactors, a number of which were actually designed and built by Argonne. Knowledge gained from the experiments at Argonne is the basis for most of the commercial nuclear reactors currently in use.

The Argonne National Laboratory is still in operation. It now has a rather wider mandate that includes basic science research, energy storage and renewable energy, environmental sustainability, and national security.

Robert's Friendship with Fermi

As they worked together on the project, Robert and Fermi developed a personal relationship. Robert reports that Fermi was a pleasant person to work with. When lunch hour came Fermi would say to his group, "Let's go and eat." They would go to the Commons, a place where both students and faculty ate. Fermi was friendly, relaxed, and easy-going. He knew where he stood. Though he did not spontaneously exhibit his ideas, he enjoyed expounding on any subject that anyone at the table chose to bring up.

Robert and Fermi both enjoyed long-distance swimming. In the late afternoons they would go swimming together in Lake Michigan, where they would swim for long distances along the shore.

The Fizzling of the Nazi Nuclear Project

Initially, on September 1, 1939 (the day the Second World War started), the Nazi military took over control of the nuclear research project in Germany. Nuclear physicists, including Werner Heisenberg, Carl Friedrich von Weizsäcker, and Otto Hahn, were brought together to meetings, and the Army Ordnance Office took over the Kaiser Wilhelm Institut für Physik in Berlin.

Otto Hahn was a pacifist, an opponent of the Nazis, and did not believe in the war. He also was strongly opposed to the persecution of Jews in Germany. He did his utmost to help support his colleagues who were suffering persecution as having some Jewish ancestry, preventing them from being sent to the front lines, and helping some to escape the country. His wife Edith collected food for Jews hiding in Berlin. He held an influential position during the war as head of the Kaiser Wilhelm Institut für Chemie, and thus had some power to influence the course of the nuclear research projects. Later, after the dropping of the two atomic bombs on Japan, he was devastated by the feeling that his early discovery of fission had lead to the death and suffering of so many people. As a pacifist, he certainly did not wish to build a bomb.

Werner Heisenberg also hated the Nazis. He was a very prominent physicist and had received the Nobel Prize in 1932. However, when he was nominated in 1935 as one of the possible replacements for Sommerfeld at the University of Munich, Heisenberg was attacked by the anti-Semitic *Deutsche Physik* group who had a bias against quantum mechanics and relativity and hated Einstein. In the July 15, 1937 issue of the SS weekly newspaper *Das Schwarze Korps* an article was published calling Heisenberg a "White Jew" (i.e., an Aryan who acts like a Jew) who should be made to "disappear."

Heisenberg's mother was a socialite in Munich with many connections. She visited Himmler's mother, whom she knew because her father and Himmler's father were both members of the same Bavarian hiking club, and asked her to have Himmler tell the SS to "give Werner a break." (Note that Heinrich Himmler was Reichsführer of the SS, a military commander, and leading member of the Nazi Party. As Chief of the German Police and later Minister of the Interior, he oversaw all internal police and security forces. This included the Gestapo — the Secret State Police — and he was one of the people most directly responsible for the Holocaust.) Himmler sent a letter to the SS saying that Germany could not afford to lose Heisenberg, and a letter to Heisenberg saying that this was "on the recommendation of his family" and that Heisenberg should make a distinction between professional physics results and the personal attitudes of the scientists.

Neither Heisenberg nor the other nominated physicists got the disputed position at the University of Munich. Eventually Sommerfeld was replaced by Wilhelm Müller, a political appointee who was not a theoretical physicist and had never published in a physics journal. However, it is noteworthy that during the SS investigation of Heisenberg, the three physicists who participated became supporters of Heisenberg, as well as of his position opposing the ideology of the *Deutsche Physik* movement in theoretical physics and academia.

Heisenberg was able to use his position as a leading scientist on the nuclear project to influence its direction to some extent. He did not wish the Nazis's to get the power that an atomic bomb would bring. However, he had to carry out a balancing act so as not to appear suspicious to his bosses in the Nazi party, even while avoiding being too successful in the nuclear bomb project. There was a well-known personal and professional animosity between Kurt Diebner and Heisenberg's "inner circle" (Heisenberg, Carl Friedrich von Weizsäcker, and Karl Wirtz) — and Diebner was the administrative director of the Kaiser Wilhelm Institut für Physik under the direction of the Army Ordnance Office.

Von Weizsäcker later stated that Heisenberg, Karl Wirtz, and he had a private agreement to study nuclear fission to the fullest extent in order to decide themselves how to proceed with its technical application. They wanted to know if a nuclear chain reaction was possible. Von Weizsäcker said, "There was no conspiracy, not even in our small three-men-circle, with certainty not to make the bomb. Just as little, there was no passion to make the bomb." He said that it was fortunate that the German war economy was unable to mobilize the necessary resources to build an atomic bomb.

In a report based on additional documents from Russian archives, historian Mark Walker concluded that "in comparison with Diebner [and] Gerlach ... Heisenberg and finally von Weizsäcker did obviously not use all power they commanded to provide the National Socialists with nuclear weapons."

(Note that some of the above information was also supplied to me via personal communications from Peter Biermann, whose family had been close to the Heisenberg family.)

When it became clear in January 1942 that the nuclear energy project would not make a decisive contribution to the war effort in the relatively near term, the army relinquished direct control of the project. It was split up between nine institutes where the directors set the research agendas (although some military funding continued). Two of these institutes were controlled, respectively, by Otto Hahn and Werner Heisenberg — Heisenberg became head of the Kaiser Wilhelm Institut für Physik in July 1942. After this point the nuclear research project essentially fizzled and the number of scientists involved dwindled, although of course the Americans did not know this. It is ironic that this was the period when the American atomic bomb project was taking off.

After the war it was discovered that the Nazi project had never built a working nuclear reactor, and none of the German scientists had calculated the critical mass of uranium-235 or plutonium-239 that would be required for a bomb. (This was demonstrated by conversations recorded in the Farm Hall Tapes of 1945, which were declassified 47 years later in 1992.)

The 1962 White House Reception Commemorating Chicago Pile-1

In the fall of 1962, President John F. Kennedy invited the scientists who had created the world's first nuclear reactor (Chicago Pile-1) to a reception at the White House, to celebrate the 20th anniversary of that event.

Unfortunately, Enrico Fermi had died of cancer years before, in 1954. In the end, out of roughly 50 scientists who had contributed to the pile's creation, 37 were present at the reception at the White House on November 27, 1962.

Robert Christy was among those who were present. Several letters concerning this event survive in his files, as well as items from an Atomic Energy Commission (AEC) "souvenir" package with photographs and the text of the press release describing the event and listing the attendees. A commemorative medallion was also presented to each of the participants.

E. I. DU PONT DE NEMOURS & COMPANY
WILMINGTON 98, DEL.

October 29, 1962

Dear Dr. Christy:

Norm Hilberry has been kind enough to sound out on my behalf those who were present on that famous occasion on December 2 twenty years ago as to their willingness to come to a dinner I would give in Washington on November 28. Norm tells me that a large representation will be able to come, which pleases me very much.

This letter then is my formal invitation to you to come to dinner at 6:30 in the Pan American Room, Hotel Mayflower, Washington, D. C., on Wednesday, November 28. My notion is to avoid all formality and simply to have a pleasant and private reunion to remember together a very inspiring occasion. Dress, of course, will be informal.

Will you please let me know if you can join me.

Sincerely,

Chairman

Dr. R. F. Christy
Kellogg Radiation Laboratory
California Institute of Technology
Pasadena, California

An invitation to attend a private reunion dinner, in contrast to the public events celebrating the 20th anniversary of Chicago Pile-1

The commemorative medallion celebrating the men who created the first nuclear reactor

The White House reception on November 27, 1962 (AEC photo)

Below is the copy sent to Robert Christy of the AEC press release:

```
                            UNITED STATES            REC___ED
                     ATOMIC ENERGY COMMISSION     DEC 4  1962
                         Washington 25, D. C.
```

                    20TH ANNIVERSARY OF NUCLEAR PROGRESS

No.   E-435   FOR IMMEDIATE RELEASE
Tel. HAzelwood 7-7831   (Tuesday, November 27, 1962)
      Ext. 3446

                    SCIENTISTS WHO BUILT FIRST NUCLEAR
                    REACTOR VISIT PRESIDENT KENNEDY

Many of the scientists who worked with the late Enrico Fermi
at Chicago where the first controlled nuclear chain reaction was
achieved on December 2, 1942 -- 20 years ago -- were received today
by President Kennedy at the White House.

Approximately 35 of the original "Fermi team" of 50 scientists,
engineers and technicians were in the group visiting the White House.
Some were accompanied by their wives. Most of them still are active
in nuclear science and engineering.

During the visit, Chairman Glenn T. Seaborg of the U.S. Atomic
Energy Commission, displayed for the President an historic souvenir
of the Fermi reactor which is to be given to the Smithsonian Institution for use in the Hall of Nuclear Energy in the new Smithsonian Building.

The artifact is a plexiglas model of the first reactor in which
is embedded a one-inch cube of uranium that was part of the original

                              (more)

- 2 -

fuel. Built at the Oak Ridge National Laboratory, the model is in-
scribed as follows: "FUEL FROM WORLD'S FIRST NUCLEAR REACTOR,
DECEMBER 2, 1942, STAGG FIELD STADIUM, CHICAGO, ILL."
   The members of the group visiting the White House were:

Dr. Harold M. Agnew
Professor Samuel K. Allison
Hugh M. Barton, Jr., and son
Thomas Brill
Dr. R. F. Christy
Richard J. Fox
Stewart Fox
Dr. Carl C. Gamertsfelder
Dr. Alvin C. Graves
Dr. David L. Hill and Mrs. Hill
Dr. Norman Hilberry
Robert E. Johnson
William H. Hinch
W. R. Kanne
P. G. Koontz
Harold V. Lichtenberger
George M. Maronde
Dr. Leona Woods Marshall (Mrs.)

Anthony J. Matz and Mrs. Matz
George Miller
George D. Monk and Mrs. Monk
Dr. Henry W. Newson and
 Mrs. Newson
Robert G. Nobles
Warren E. Nyer and Mrs. Nyer
W. P. Overbeck and Mrs. Overbeck
Howard Parsons and Mrs. Parsons
Dr. Gerard S. Pawlicki
Theodore Petry
David R. Rudolph and Mrs. Rudolph
Leon Sayvetz
Dr. Frank H. Spedding
Dr. William J. Sturm
R. J. Watts
George L. Weil
Dr. Marvin H. Wilkening and
 Mrs. Wilkening
Dr. Volney C. Wilson
Dr. Walter H. Zinn

- 30 -

11/27/62

Achieving the Rare

Below is a letter from Duncan Clark of the AEC, written to each of the Fermi team members, with a photograph of a lucite model of the first nuclear reactor:

UNITED STATES
ATOMIC ENERGY COMMISSION
WASHINGTON 25, D.C.

January 10, 1963

Members of the Fermi team:

First, may I thank you for your help in making the 20th anniversary of the Fermi reactor a truly national observance.

As was promised you at the December 1 luncheon at Chicago, we have assembled this small "souvenir" package. It contains the text of the remarks made at your reception at the White House on November 27, one of the photographs made at that time, a picture of the first pile artifact fabricated at Oak Ridge National Laboratory, a copy of "The First Pile," the major background article prepared by our Public Information staff for the 20th anniversary, and a copy of the educational poster, of which 30,000 were distributed.

We understand that as a group, already you are discussing possibilities for some type of a permanent memorial to be completed possibly by the 25th anniversary in 1967. I am asking Charter Heslep to maintain an informal liaison with you on the public information aspects of any plans you may develop.

Sincerely yours,

Duncan Clark, Director
Division of Public Information

Lucite model of Chicago Pile-1, the world's first atomic reactor, containing one of the fuel pellets from the original reactor (AEC photo)

Chapter 6

Designing the First Atomic Bomb

It was early in 1943. Oppenheimer had just been appointed director of the Los Alamos branch of the Manhattan Project, and was recruiting people to work with him. He traveled to the University of Chicago and asked Robert Christy to join him, telling Robert that they would be building a nuclear weapon.

It was due to the Japanese attack on Pearl Harbor on December 7, 1942 that the U.S. had joined the allies fighting the Second World War. However, the nuclear weapons were intended to be used against Nazi Germany rather than Japan. The Allies had heard from Bohr that Hitler had commissioned Heisenberg and his colleagues in Germany to build an atomic bomb.

Robert Christy's ID photo at Los Alamos

There was plenty still to do at the University of Chicago with Fermi. The first reactor, which the Chicago group had just created, had been essentially a proof of concept and had produced only half a watt of heat. A higher-powered reactor was to be designed at the new Argonne site, to produce energy. But when Oppenheimer called, Robert went. He agreed immediately to join Oppenheimer at Los Alamos. The pay at Los Alamos was initially the same as at the University of Chicago, with small pay raises at intervals thereafter. Robert described his recruitment by Oppenheimer:

> Early in 1943, I was recruited by Robert Oppenheimer to go to the new laboratory that he was setting up in Los Alamos. My first job, actually, was to visit some of the fast neutron laboratories: the University of Minnesota, and I think the

70

University of Wisconsin, where they were doing experiments with fast neutrons, to assist in analyzing some of their data and in general to assist them in working out some of the results they were getting. I spent a couple of weeks doing that, before I actually arrived in Los Alamos.

Robert's Arrival at Los Alamos in Early 1943

Robert was one of the first to join Oppenheimer at Los Alamos. There wasn't much there. No accommodations existed at Los Alamos yet, and Robert lived in an existing dude ranch in the valley below, miles from Los Alamos. Fifty to a hundred people were soon living in these dude ranches, with buses to bring them to the Los Alamos site each day. Serber, who had been a postdoctoral fellow of Oppenheimer's at Berkeley, gave a series of lectures at Los Alamos, "The Los Alamos Primer," to acquaint the scientists with the ideas that existed at the time of how to create atomic bombs, and what the essence of the project was.

Buildings, laboratories, and living facilities were being constructed with greater speed than Robert had ever seen before or since: it was only a few weeks before housing was available at Los Alamos itself. Initially, most of the scientists were packed into a common dormitory sleeping in double-decker bunk beds, some of them out on a porch. There is a famous story of how Feynman, trying to get to his upper bunk, had to crawl over the bed of Robert and Dagmar Christy's lower bunk — which drew pointed remarks from Dagmar that Feynman never forgot. This was one of the humorous stories that Feynman enjoyed recounting in later years.

Soon there were one-bedroom duplexes available for the married scientists. Robert and Dagmar were assigned one of these just across the road from Oppenheimer's house. This was "kind of in the social area," as Robert put it. However, the stoves in the kitchens of these apartments were wood-burning ones which took half an hour to an hour before they were hot enough to cook on and really had no way to regulate the heat. The women refused to cook on these stoves — they were accustomed to instant-response gas or electric stoves. As a solution hotplates were supplied, and for the following years almost all the cooking in these apartments was done on hotplates. Turkey cooking for Thanksgiving and Christmas was a challenge. For the seven following decades, Robert treasured the portable electric oven that they had used to cook the turkey — he continued to use it each Thanksgiving.

When Robert and Dagmar's first child was imminent, and the one-bedroom apartment they inhabited at the time would be too small, they were moved to a two-bedroom apartment in a fourplex. This set of four apartments was unfortunately located out in the sticks far from the social center of Los Alamos.

In this connection, it has to be stated that nobody at Los Alamos was allowed to leave to visit relatives, even at Thanksgiving or Christmas. Los Alamos was a fenced site with military guards. All mail was read and censored. It was a secret site, and people were not allowed to communicate with anyone outside about what they were up to. They were, as Robert called it, "confined to base." However, they were allowed to go shopping in Santa Fe and to visit the Indian pueblos and the restaurants nearby, being carefully checked by the guards on each entrance and exit. A few exceptions were made. For example, Feynman got permission to make trips to visit his wife who was very ill with leukemia in a hospital in Albuquerque.

Robert's Bosses at Los Alamos: Peierls, Bethe, Oppenheimer, and Groves

Robert was first assigned to the theoretical group headed by Hans Bethe. Later, in August 1943, the group of physicists from England arrived. One of these was Rudolf Peierls, who had studied solid state physics in Zürich under Heisenberg and Pauli. He had been in Britain on a scholarship when Hitler came to power. As a German Jew he was granted leave to remain in Britain, where he worked with Bethe and with Wigner. When the war began in 1939 Peierls started working with Frisch and Chadwick on nuclear research. After coming to Los Alamos in 1943, Peierls became Robert's immediate boss as a division head under Bethe.

Robert was well acquainted with Bethe's reputation as an eminent physicist, and of course he knew Oppenheimer even better from having been his student at Berkeley. Robert had had no prior knowledge of the Manhattan Project's ultimate boss, General Leslie Groves, but during the time spent working at Los Alamos Robert came to admire both his drive and his abilities.

Two decades later, in 1964, Robert was asked to nominate someone for the Enrico Fermi Award, which was first established in 1954 for "any especially meritorious contribution to the development, use, or control of atomic energy." It was awarded to Fermi himself in 1954, and subsequent recipients included the mathematician von Neumann and the physicists Lawrence, Wigner, Seaborg, Bethe, Teller, and Oppenheimer. Robert nominated Groves, saying, "In my view, he made unique contributions to the direction and implementation of the Manhattan Project efforts towards atomic power and atomic weapons. It is unlikely that the objectives of the project would have been so quickly accomplished without his participation as director." But the 1964 Fermi Prize did not go to Groves. It was awarded instead to Hyman G. Rickover, the admiral who had directed the development of naval nuclear propulsion.

The Spy Klaus Fuchs

Klaus Fuchs, who later was discovered to be a spy for the Soviet Union, was part of the English group that arrived in August 1943. He ended up with an office next door to Robert. Nobody at Los Alamos knew that he had been a spy for the Soviets since 1941.

Only after Fuchs returned to England in 1946 was it discovered that he had passed on secrets of the atomic bomb project, starting in England and continuing in the U.S. These secrets had included plans of the fission bombs developed at Los Alamos and the early (not yet practical) designs for the hydrogen fusion bombs developed by Teller. In 1950 he was convicted in England. However, because the Soviet Union was still classified as an ally ("a friendly nation") during the war years, Fuchs could only be sentenced to a maximum of 14 years in prison. He was actually released after only nine years, and in 1959 went to Communist East Germany. There he was rewarded with several prominent positions. He was elected to the Academy of Sciences and the Socialist Unity Party of East Germany's central committee, and later was appointed deputy director of the Institute for Nuclear Research in Rossendorf, near Dresden; he received the Fatherland Order of Merit and the Order of Karl Marx. It has been suggested that a tutorial he gave to Chinese physicists may have helped them develop their own atomic bomb.

Hans Bethe once said that Klaus Fuchs was the only physicist he knew who truly changed history. However, some former Soviet scientists have stated that they were actually hampered by Fuchs' data. The head of the Soviet project, Lavrenti Beria, insisted that the Soviet bomb resemble the American plutonium bomb as much as possible, even though the Soviet scientists had discovered a number of improvements that could have led to a more efficient version of their atomic bomb.

Robert's "Instant Fame" from His "Water Boiler" Reactor Calculations

At first Robert worked on a slow neutron reactor that used enriched uranium. He developed what was later called the "water boiler," a homogeneous reactor that uses enriched uranium as fuel and ordinary water as moderator. The fuel was in the form of uranyl sulfate dissolved in the water. The spherical container holding the uranium solution was surrounded by beryllium oxide bricks. Neither beryllium nor oxygen absorbs neutrons, so many of the neutrons from the uranium would be reflected back into the "boiler." This would allow the use of a smaller amount of uranium than would otherwise be required.

The importance of this "water boiler" reactor was that it would not require much enriched uranium. Since enriched uranium was only being produced very slowly, the "water boiler" would allow the Los Alamos teams to get experience with critical masses well before other types of tests could be performed. In addition, unlike the Chicago reactor, the water boiler would be similar in size to an actual bomb. Therefore the calculations for it would more closely resemble the calculations required for a bomb, allowing a better test of the theory. The idea was that any chain reaction in the "water boiler" would limit itself because the water would start to boil if the chain reaction got to be too strong. The resulting bubbles would cause the water to expand, and the expansion would stop the chain-reaction. It would be a *self-limiting* chain-reaction, and therefore very safe.

Robert calculated the properties of the neutron reflector and the water, and the critical mass of uranium that would be needed to sustain a chain reaction. By the time enough U-235 had been accumulated to test this type of reactor some new cross sections were available, and Robert hastily revised his calculations. He achieved almost instant fame at Los Alamos when it turned out that the amount that he calculated was within a very few percent of the actual amount that they found was needed to go critical. As Robert put it, "So I acquired tremendous fame — here is a theoretical physicist who calculated the right number! ... Now I will confess that anytime you hit something within a percent, it's largely luck. But I didn't go around telling people it was luck. So that was a triumph."

Criticality was achieved in this "water boiler" reactor on May 9, 1944. A higher-powered version with cooling and shielding was then designed and built to serve as a neutron source. Among other things, this reactor allowed the Los Alamos scientists to measure neutron absorption cross sections of materials that they were considering using. This second "water boiler" reactor continued in service until 1950, when an even higher-powered version was constructed.

The One-Dollar Patents at Los Alamos

Robert described the Los Alamos policy for patents:

> The policy at Los Alamos was to patent all discoveries, things that we did that were new, simply to forestall anyone else patenting it and then charging the government royalties. So everything that was done there of such a kind was patented. I received a patent on sketches that I made for this water-boiler reactor, and also the calculations that I made for the amount of uranium-235 that would be needed. I received several patents by the U.S. Government. For these patents, the government paid one dollar — but in fact we never even received the dollar. This was standard.

The "water boiler" patent is included at the end of this chapter. A patent was also received for the design of the first atomic bomb exploded in the Trinity test at Alamogordo (and later at Nagasaki).

Understanding Large Explosions such as the Volcano Krakatoa

Robert next worked on understanding atomic explosions and their effects. To gain understanding of the effects to be expected, he studied large natural explosions. He read as much as he could on the explosion of Krakatoa, the Indonesian volcano that exploded in 1883 with a force equivalent to 200 megatons of TNT (13,000 times as powerful as the Hiroshima bomb). This explosion destroyed the island, killing 40,000 people, and was heard as much as 3,000 miles away.

Robert recalled, "I remember calculating the effects of setting off an atomic bomb underwater — what kind of shock waves, what kind of waves, etc., would be made." For this work on explosions Robert considered several different possible configurations for an atomic bomb.

Robert's Work on the Implosion Bomb: the "Christy Gadget"

Robert described some of the work on the design of the implosion bomb, including his crucial contribution:

One project I worked on was to try to calculate the properties of materials like uranium [and plutonium] under exceedingly high pressure, which might be achieved in an implosion bomb. I had to try to develop a theory that would tell how much uranium might be compressed under these circumstances. This was kind of an extension of anything that had been done before, since no one had ever tried to evaluate the properties of materials under those circumstances. That was one task that I had.

Then, about this time, the implosion work became very important, and I was assigned under Rudolf Peierls to work on the implosion bomb. I worked on that for some time. In fact, for the rest of the time at Los Alamos, my primary occupation was working on the implosion bomb, although I did some side jobs. But in the implosion calculations, I eventually came up with the idea of the bomb that was finally made and tested at Alamogordo and dropped at Nagasaki. This was a modification of the designs that we had been working on — a modification that was adopted by the laboratory, because the problem with the implosion bomb was that it had to have a highly symmetrical implosion in order to work, and people were very much concerned that it would not be symmetrical, and therefore the bomb would be a fizzle and this would be a disaster (if a bomb was dropped that didn't work) and General Groves might even be demoted. Anyway, we were very concerned that the bomb should work.

The idea I had was to eliminate the hole in the middle of the implosion bomb and make it a solid sphere [of plutonium] that would be compressed by the high explosive. This eliminated the possibility of the material not arriving at the center symmetrically, because it would start off symmetrically, and would have nowhere else to go. This suggestion was adopted by the laboratory, and eventually became the basis for the Alamogordo bomb and the Nagasaki bomb. I received a patent on that too, simply for the same reason I mentioned before [to forestall anyone else patenting it and then charging the government royalties]. That's why it was called the "Christy Gadget."

At Los Alamos the two designs for the atomic bomb were finally ready. The first atomic explosion was the test in the desert at Alamogordo, New Mexico; this bomb's design was the "Christy Gadget" described above. This design used a sub-critical sphere of plutonium-239 imploded by high explosives to compress it to a high enough density that it became super-critical.

The critical mass is the minimum mass needed to sustain a nuclear chain reaction that will grow instead of petering out. To get an explosion (rather than just getting the plutonium to heat up and melt), the critical mass must be assembled very quickly. There was concern that a hollow spherical shell would crumple when imploded and therefore fizzle instead of explode. As he described above, Robert's crucial contribution was to suggest that a solid sub-critical sphere be used rather than a hollow shell. The critical mass of plutonium-239 that is typically quoted is approximately 10 kg for a solid sphere that has not been compressed; but due to the compression from the explosives this bomb actually needed only 6 kg. This amount could be produced in about a week in the nuclear reactor at Hanford, Washington.

Why the "Christy Gadget" and Not the "Christy Bomb"?

Robert recalled,

> It was called "gadget" rather than "bomb" because we didn't use terms like "bomb" and so forth; these terms might be used in ordinary conversation, that eventually could attract the attention of the people in the neighborhood, if you were walking along the street in Santa Fe and talking about bombs. So there were code names used, just to avoid that. One of the stories that we tried to put forward was that we were working on some new kind of submarine up there at Los Alamos. But anyway, there were stories put out to try to explain what was going on without revealing the actual work.

The Trinity Test for the Implosion Design

Robert Christy's implosion design needed to be tested: as he told us,

> I would say that the final design that I made had convinced people that it would work, although there were still people who were worried about it because the whole concept of implosion was rather a radical concept, and people were not as certain about it as they were about the "gun assembly" method.

The test took place at Alamogordo, on July 16, 1945.

It is hard to believe, but the core of this powerful bomb, the nuclear material, was actually transported from the secret assembly site at Los Alamos to the test site in the desert not in an Army vehicle but in the personal station wagon of Bob Bacher, the head of the experimental group of the bomb project. Robert was one of those who observed the test from about 25 miles away. Robert described it in his 2006 interview with Sara Lippincott:

> We were assigned the kind of glass that is used in a welder's mask. Basically you can't see through it, except for brilliant light like a welder's arc. So we were told to hold this up in front of our eyes. And we did. And even with that, you see in your peripheral vision the whole world light up like the sun is out. But through the glass you can see the actual explosion going off. It was awe-inspiring. It just grew bigger and bigger, and it turned purple.

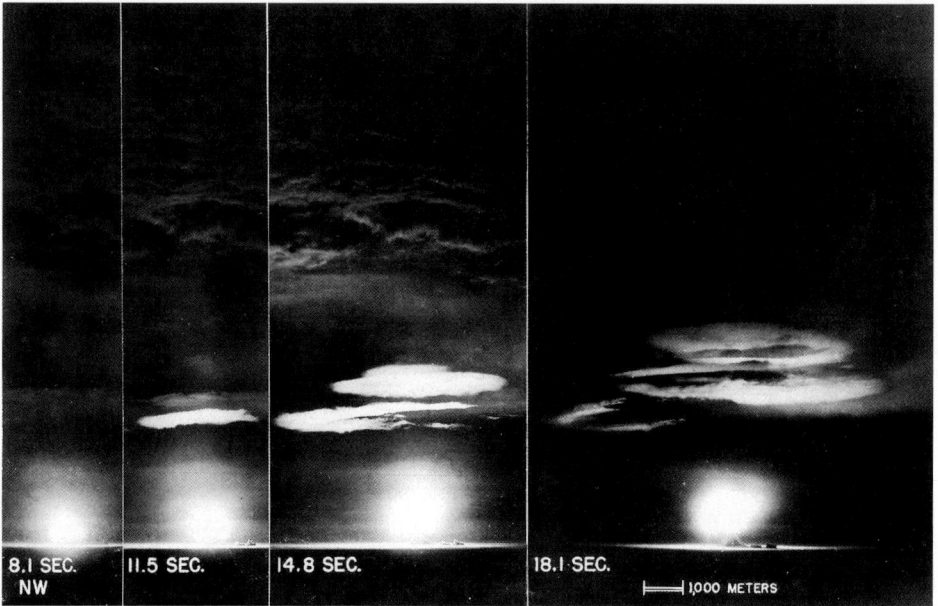

8.I SEC.   II.5 SEC.   I4.8 SEC.   18.I SEC.
NW   ⊢——⊣ I,OOO METERS

Photos of the Alamogordo test at four stages, from 8.1 to 18.1 seconds after the explosion

I5.O SEC.   ⊢——⊣ IOO METERS
N

Close-up of the Alamogordo test, 15 seconds after the explosion

    The purple was an interesting thing which I certainly hadn't anticipated. But
it was in this ball. The debris was intensely radioactive, and it was sending out
beta particles and gamma rays in all directions, and those ionized the air. So the
air around this ball emitted a bluish glow which comes from ionized air. It was

most fantastic, to see this thing going up and swirling around and eventually cooling off to the point where it was no longer visible.

The first reaction to the success of this atomic bomb was joy that their work had not been in vain. Brigadier General Thomas Farrell turned to his immediate superior General Leslie Groves, the head of the Manhattan Project, and said, "The war is over." Groves replied, "Yes, just as soon as we drop one or two of these things on Japan." Soon, however, the scientists began to fear the consequences of this new weapon.

First Attempts to Prevent the Actual Use of Atomic Bombs

Nazi Germany had surrendered on May 7, 1945, and therefore the initial motivation for the development of the atomic bomb was no longer relevant. However, the war against Japan continued in the Pacific. In Robert's words:

> After that was still a very tense period, because the war was going on. The marines were fighting on one island after another in the Pacific. There were many casualties, so that there was a very intense and bloody war going on in the Pacific, and no one was relaxing until that war was over. As far as the bombs went, the next step was to transport materials and prepare them for dropping on Japan. So there was a very intense activity, but involving a smaller number of people.
>
> At Chicago, Szilard and others were trying to petition the President and the government to make a demonstration of the bombs rather than dropping them on cities. My feeling is that the reason for that political activity at Chicago was that at Chicago their principal work had been accomplished already in December 1942. After that they did some more work there at Chicago, but their mission had been accomplished. The [plutonium] production reactor design went to the Du Pont Company, with advice from the people at Chicago, and the production reactors were made at Hanford, Washington, so that Chicago was left without very much to do. If you leave scientists without very much to do, they begin to worry about problems; and so they worried about the use of the bombs at a stage before anyone else did, because they were not otherwise occupied. They developed the idea that it should be publicly demonstrated, not used immediately in warfare.

(Note that while 68 scientists from the University of Chicago's Metallurgical Lab signed Szilard's petition, it never actually reached the President, because General Groves strongly opposed it on the grounds that such a petition would breach security and expose the existence of the atomic bomb. The petition was not declassified and made public until 1961.)

The proposals for how to deal with the bomb were considered by various committees of the government, in which I was not involved. In fact, I think the only person at Los Alamos who was involved was Robert Oppenheimer. Fermi may have been involved, I think, and [Isidor] Rabi. A few of the senior people were asked their views. They advised the government, and then the government went ahead and made their decisions. I don't know the chain of that, who made what decision, but ultimately responsibility for that rested with the President.

I believe, from what I have read, that President Truman — who took over after Roosevelt's death — President Truman's concern was that every day, every week, thousands of people were losing their lives in the war, and I believe his concern was that he could not face the American public if he had not used every means available to stop this bloody warfare as soon as possible. I think very few people have questioned the correctness of that decision, given his responsibilities and given the way the war was going. Actually many people in Japan were dying every day, including many civilians. There were constant air raids on Japan, and fire raids had been used there which wiped out large portions of many big cities. In fact most of the big cities in Japan, including Tokyo, were devastated by fire. These fire raids actually caused many more casualties in Japan than the atomic bombs did, but they did not have the same dramatic effect as the atomic bombs. I think it can be contended that the atomic bombs saved not only American lives but also Japanese lives.

The war in the Pacific had become very bitter, with huge losses of lives on both sides. In February 1945 at Iwo Jima, about 7,000 American marines were killed, with 20,000 wounded; 20,000 Japanese soldiers were killed. In the Battle of Okinawa (April-June 1945), there were over 80,000 American casualties (killed or wounded), and 108,000 Japanese troops died, along with thousands of Japanese civilians. Much higher losses — millions of lives — were projected if the U.S. had to invade Japan. It has been suggested that President Truman might have been impeached if he had not made use of the new atomic bombs.

Firebombing vs. Atomic Bombs

The firebombing had been initiated because the Japanese had intentionally decentralized 90 percent of their war-related production into small subcontractor workshops in civilian districts, making the Japanese war industry largely immune to conventional precision bombing with high explosives. Air Force General Curtis LeMay ordered massive incendiary bombing of 64 Japanese cities between March and August 1945. This included the massive Tokyo firebombing on the night of March 9-10, which incinerated 16 square miles of the city and is estimated to have killed 100,000 civilians.

The total estimated casualties from the firebombing raids are 500,000 civilian deaths, with 5,000,000 left homeless. This is more than twice the estimated casualties from the atomic bombs of Hiroshima and Nagasaki combined. The immediate death toll at Hiroshima was estimated to be between 70,000 and 80,000, with a total of 90,000 to 160,000 when one included those who later died from their injuries. For Nagasaki, the immediate death toll was estimated to be 40,000 to 75,000, with total deaths of about 80,000.

Groves' Motivation for Dropping the Bombs

In a letter to Robert Christy dated 10-1-1998, British historian A. Murphy quoted part of a letter/cable dated 6-3-1945 from the British ambassador Lord Halifax to Sir John Anderson, describing a visit from General Groves:

> In speaking of the finance of the project Groves referred to some of his difficulties with Congressmen and said that immediately after the war was over they would certainly be faced with a very searching investigation by Congress. He himself would probably be the chief witness. ... Many of his actions which might have seemed to us rather unreasonable were inspired by his concern over this future Congress investigation. ... Groves emphasized the importance for the future of the project of using at least one bomb during the present war. He hoped on Tokyo. He was sure that this was the only kind of demonstration of the effectiveness of the new weapon which would convince the United States Congress and the public of its importance or persuade them to vote money for the continuance of the work; trials in the desert or on an island might be sufficient for scientists, the Army etc.; but they would not appeal to the public.

It is estimated that the Manhattan Project cost the U.S. taxpayers two billion dollars, an immense amount of money at that time (equivalent to 40 or 50 billion dollars today). From the letter above, it is apparent that General Groves was concerned that the project might be terminated unless there was a very visible sign of success to justify the expense. This provided an additional motivation to push for the use of the bomb, besides trying to put an immediate end to the war.

The historian Murphy came to the conclusion that the war with Japan could have been ended at least a month earlier than it actually did, even without the use of the atomic bomb, if the Allies had been prepared to guarantee that the Japanese Emperor would be allowed to continue his tenure and not be tried for war crimes. Murphy concluded that the Allied demand for an unconditional surrender was what delayed the surrender until the atomic bombs were dropped. One should note that in the end the Emperor was not deposed, nor was he tried for war crimes.

The Use of the Two Atomic Bombs

The first atomic bomb actually dropped on a city was at Hiroshima in Japan on August 6, 1945; it was nicknamed the "Little Boy." This was not a plutonium implosion bomb but rather was of the other Los Alamos design, which used nearly-pure uranium-235 in two sub-critical pieces. One of these pieces was a projectile that was fired at the other piece (down a gun-like tube) in order to assemble the critical mass of uranium-235 rapidly enough. This was therefore called a "gun-type" design. The critical mass of pure uranium-235 is usually quoted as about 52 kg; to make this bomb required 64 kg of uranium-235. It is noteworthy that this design had not been tested before it was dropped on Hiroshima, because it would have taken several more months to produce enough uranium-235 to build another such bomb (the diffusion plant at Oak Ridge, Tennessee was the only source of this material), and because there was more confidence that it would work. Robert Christy described this bomb design:

> The gun assembly was a simple concept. The idea of an atomic bomb is to take two portions of fissionable material — in the case of the Hiroshima bomb it was uranium-235, which had to be of course separated from U-238 in order to get the pure U-235. The bomb is going to explode if you have more than a critical mass assembled together at one time, and you don't have a long time to do it because there are stray neutrons floating around in the air all the time (coming from cosmic rays), and if a stray neutron hits the material when it is only slightly over critical, then the bomb will go off but with a very low yield and it would be called a fizzle.
>
> The idea, then, is to assemble the two portions of U-235 in a time small enough that no stray neutrons are apt to be there. The way proposed to do this was fairly direct, and that is to have a target consisting of one slightly subcritical piece of U-235 and a bullet or actually a shell consisting of the other slightly subcritical piece of U-235, and the bullet or shell is fired from a gun into the target. When they finally merge in the target, there is one single piece that is supercritical. This merging has to be done fairly quickly, and that's why a gun was used, simply to avoid stray neutrons. A gun brought the materials together in a time like milliseconds, and in that time there was very little chance of stray neutrons being around. The halves could thus be assembled in time to make a combined object which could then be initiated by an initiator, to go off.

Once initiated the chain reaction builds up exponentially, one making two making four making eight making sixteen, and after about eighty generations, that is eighty steps of this kind, the number of neutrons and the number of fissions becomes comparable to the number of atoms of uranium present, and then you have a very large explosion. This takes about eighty generations, or eighty steps, and the time involved per step in fast neutron reactions is something like what was called a "shake" (that was a term invented at Los Alamos to describe an interval of time of 10^{-8} seconds, i.e., ten nanoseconds). Eighty steps of that kind took around one microsecond. This was the time it took for the reaction to develop, and the idea was to keep it in the assembled state during this period of a few microseconds, and that way the reaction goes off.

Robert Christy had been scheduled to fly in the airplane with this bomb when it was dropped; he was to be a key scientific observer for the bomb. However, he had pneumonia at the time and was removed from the mission. This bomb had never been tested — it was not known *exactly* how powerful it might be — but its design was straightforward and its performance was almost guaranteed. Thus an observer was not crucial in this case.

The Japanese generals did not immediately surrender, so a second atomic bomb was dropped three days later on Nagasaki on August 9, 1945. This one was nicknamed the "Fat Man." It was of the "Christy Gadget" implosion design, using plutonium-239. There was no need for Robert Christy to be on board to observe this bomb since the design had already been tested at Alamogordo.

Robert Christy in 2006 (at age 90), standing beside a replica of the "Fat Man" bomb in the Los Alamos museum

After the War Had Ended

Robert spoke of the reactions of the researchers at Los Alamos when the war was over:

The real jubilation occurred at Los Alamos when the Japanese surrendered, because that was the end. After the Japanese surrendered, it was remarkable that although everyone was still there at Los Alamos for the next six months or so, almost nothing got done in the way of work on the project. The project had to be re-defined. But the fact is that people stopped working and started discussing all sorts of questions having to do with atomic energy, atomic weapons, international control, and so forth. The people did not continue to work on developing bombs after that, essentially.

On August 30, 1945 Robert became one of the founding members of the Association of Los Alamos Scientists (ALAS), which had the goal of supporting only peaceful uses of nuclear energy (as described in Chapter 7).

Robert's Memories of His Colleagues at Los Alamos

Robert interacted with a considerable number of scientists at Los Alamos, and had gotten to know others with whom he did not work directly. Many had originally come from Europe, including Germany. Robert recalled:

One of the most infamous, of course, was Klaus Fuchs. He had one kind of calculation under Peierls, and I had another; our offices were just across the hall. He had a parallel position to me — he didn't work for me, and I didn't work for him, but we both worked with Peierls. He [Klaus Fuchs] was noted ultimately, of course, as being a spy who was conveying information to the Soviet Union.

I worked with Hans Bethe, who was the division chairman at the time. Victor Weisskopf was a section leader, carrying out some of the calculations in theory for the division. Dick Feynman was there. Bob Serber was there. I'm sure there were a number of others, but it was a large laboratory of course, and I knew many people there.

Fermi came to Los Alamos later, perhaps in late 1944 or something. He was busy carrying out experiments with the chain-reaction in Chicago until then. When he came, the lab was all set up so that he was not made a part of the teams that were making the bomb. He was doing some parallel experiments, and he also had a very important role advising the director — he was in charge of some parallel work on a small reactor, as well as advising the director.

Robert did not work with Teller; in fact, very few people did:

At Los Alamos, he worked independently. The main project was aimed at making the atomic bombs, and the theoretical work for that was under Hans Bethe. Edward [Teller] did not, I think, want to work under Hans Bethe's direction, so he was given an independent assignment and most of his interests and effort went into thinking about the so-called "super," namely, the bomb made with deuterium [the hydrogen bomb]. That did not essentially bear on the development of atomic bomb, and so his work was quite separate from that of most of the Project. He had a small group of, I think, two or three people.

Robert's Memories of Visitors von Neumann and Bohr

Robert described the visits to Los Alamos of two eminent scientists, the mathematician John von Neumann and the physicist Niels Bohr:

We had a number of people who would visit there from time to time. One of them was a very famous mathematician, John von Neumann, who turned out to be a most remarkable person. He was a very good mathematician and he impressed even mathematicians, but he could also talk to physicists and explain things in ways that made them absolutely clear, very simple. So he played an important role in the whole implosion work by helping set up the theory and the calculations in a way that they could be carried out.

Another visitor who came from time to time (but not very often) was Neils Bohr. He played a role as more of a consultant and critic. I remember at one stage, late in the development of the implosion bomb, we were trying to plan the initiator, a small device placed at the center of the bomb that would be sure to give off quite a number of neutrons as soon as the explosion wave reached the center. If it failed to give off neutrons then of course the bomb might fail, because the bomb depends on having neutrons present to initiate the reaction at the time when the conditions are the most favorable. In the case of the implosion bomb, that is a very short period of time when the materials are compressed and conditions are highly favorable: probably less than a microsecond. So the design of the initiator was important. We had developed a design which we called the "urchin" (named after the sea urchin), and we were debating whether this would be satisfactory. At one time Bohr came to my office, and we argued for some period of time. He was trying to be the critic and find out any flaws in the design, and I was trying to maintain that it was a sound design and that it was guaranteed to work. That was the kind of role he played. I'm sure he visited many other people, to question them about the work they were doing and to try to argue with them and see whether or not there were any holes in what they were thinking of.

One of Bohr's main roles I was vaguely aware of, but not involved in. He and some other visitors consulted with the director and a few others of the senior members of the laboratory, primarily on political questions of international control and how to deal with the problems of atomic energy after the war, and things like that. As I said, I was not involved in that activity, but he was.

Recreational Activities at Los Alamos

There was a fair amount of partying in peoples' apartments and homes, and at Fuller Lodge there were sometimes dances, with a lot of gaiety and of course some drinking. Many of the wives were working in various positions at Los Alamos.

Bob Christy's 1937 Ford – 2 door Sedan. I bought it ~ 1946.

Aggie Naranjo.

The car in the photo above is the one that Robert bought in 1945 and owned for one year while at Los Alamos, selling it (to Aggie Naranjo) in 1946 before he left Los Alamos. It was also useful for his colleagues' errands and for ski trips.

While at Los Alamos, Robert first began to ski. He wished to accompany the European physicists such as Hans Bethe, Victor Weisskopf, and Enrico Fermi who had learned to ski in Europe. Basically their method just consisted of climbing up a mountain and skiing down.   According to Robert, this was not very elegant; the only elegant skier among them was George Kistiakowsky, a chemist who had emigrated from Russia after the Communist Revolution.

Robert had obtained his skis and ski boots from a store in Santa Fe. However, they couldn't fit his size 13 feet: the largest boots they had were size 12. So he learned to ski in boots a size too small, with his big toes curled up.

Oppenheimer was a horseback rider, and owned a small horse ranch near Los Alamos. He was accustomed to riding for days all the way to Colorado, sleeping out. He would invite other physicists to ride with him around his ranch while doing physics on horseback. When the Manhattan Project had needed an isolated site far from possible spies, Oppenheimer had suggested the Indian pueblo area of New Mexico near his horse ranch. There had been a boys' school at Los Alamos called the Los Alamos Ranch School, where horsemanship was part of the curriculum. This school was taken over by the army for the Manhattan Project.

Oppenheimer gave one of his horses to be shared by Robert Christy and Lloyd Williams (who was the son of a physicist at Los Alamos, Johnny

Williams). Robert and Lloyd had joking discussions about who owned the front half of the horse which had to be fed, and who owned the back half from which the manure had to be removed. Nobody demonstrated to Robert how to ride: he had to figure it out on his own. One day near the beginning he wanted to turn one way, and leaned that way to try to tell the horse so. But the horse decided to go the other way and they nearly parted company. It was only by such trial and error incidents that he learned how to direct the horse.

There was also a certain amount of hiking done around Los Alamos. The European scientists were accustomed to climbing mountains. One favorite was to hike to where the beautiful Valles Caldera could be seen — it was too far to reach in the time they had for recreation.

Robert had always loved the mountains, but he also grew to love the desert. Years later, in a letter of October 31, 1970, he wrote to me:

> I want you also to see and learn to enjoy the desert. It will be for you an entirely new experience. In many ways it can seem harsh and barren at first, but then you see the beauty — of the light, the shadows, the shapes of the mountains, the rocks, the landscape, the colour of the rocks and soil, and when you look more closely, sparse but also very interesting vegetation. The season to enjoy it is fall (now), winter, or spring. In summer it is too hot. When I got up this morning the rising sun was illuminating with a somewhat unreal rosy light the mountain that rises a mile or two to the west. As the day wears on the light changes and so does the landscape.

It is noteworthy that the women at Los Alamos seldom participated in the skiing, horseback riding, or hiking. Many of the women worked at Los Alamos in various roles, but some were busy having — and taking care of — babies.

Robert's first son, Thomas Edward Christy ("Ted"), was born on August 30, 1944. It is amusing that the work was so secret that Ted's birthplace was listed only as "P.O. Box 1663, New Mexico." This was the only official address of all of the people at Los Alamos (several thousand people by the end of the Manhattan Project).

Robert's second son, Peter Robert Christy, was born in Chicago on May 4, 1946, nine and a half months after the first "Christy Gadget" test explosion at Alamogordo. Coincidentally, Peter Christy's first child, a daughter called Taylor Elizabeth ("Tess"), was born on July 16, 1991, the 46th anniversary of the test explosion at Alamogordo.

The Patent for the "Water Boiler" Reactor

UNITED STATES
ATOMIC ENERGY COMMISSION
CHICAGO OPERATIONS OFFICE
9800 SOUTH CASS AVENUE
ARGONNE, ILLINOIS

August 31, 1961

Mr. Robert F. Christy
W. K. Kellogg Radiation Laboratory
California Institute of Technology
Pasadena, California

Subject: U. S. PATENT NO. 2,986,510; AEC CASE NO. S-98

Dear Mr. Christy:

I am pleased to enclose for your files a copy of the
above patent.

I want to thank you for the courtesy and assistance
which you extended to members of my staff in connection
with the preparation and prosecution of the patent
application.

Very truly yours,

George H. Lee, Chief
Chicago Patent Group

Enclosure:
Patent No. 2,986,510

May 30, 1961   E. P. WIGNER ET AL   **2,986,510**

MASSIVE LEAKAGE IRRADIATOR

Filed May 14, 1946   2 Sheets—Sheet 1

FIG.1.

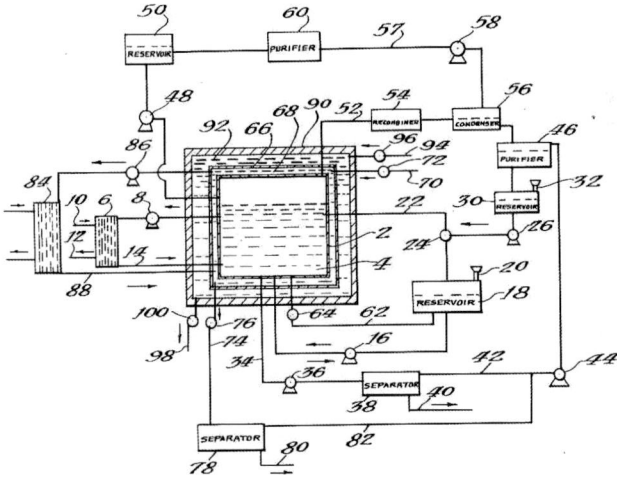

Inventors:
Leo Szilard
Eugene P. Wigner
Robert F. Christy
Francis L. Friedman
By: Robert A. Somerville
Attorney.

Witnesses:
Herbert E. Metcalf
Walter L. Schlegel, Jr.

May 30, 1961   E. P. WIGNER ET AL   **2,986,510**

MASSIVE LEAKAGE IRRADIATOR

Filed May 14, 1946   2 Sheets—Sheet 2

FIG.2.

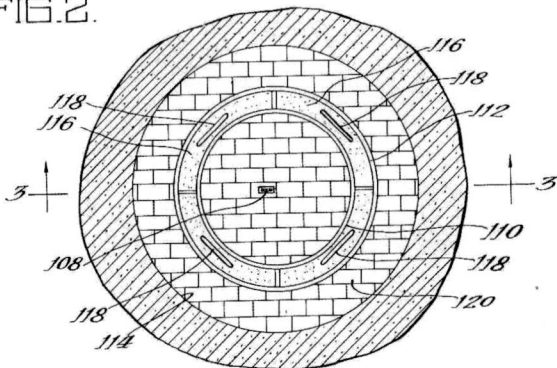

FIG.3.

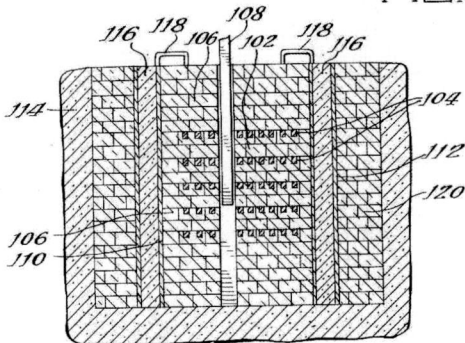

Witnesses:

Herbert E. Metcalf

Walter L. Schlegel, Jr.

Inventors:
Leo Szilard
Eugene P. Wigner
Robert F. Christy
Francis L. Friedman

By: *Robert A. [signature]*
                              Attorney

United States Patent Office

2,986,510

Patented May 30, 1961

1

2,986,510

MASSIVE LEAKAGE IRRADIATOR

Eugene P. Wigner, Princeton, N.J., Leo Szilard, Chicago, Ill., Robert F. Christy, Santa Fe, N. Mex., and Francis Lee Friedman, Chicago, Ill., assignors to the United States of America as represented by the United States Atomic Energy Commission

Filed May 14, 1946, Ser. No. 669,524

1 Claim. (Cl. 204—193.2)

This invention relates to neutronic reactors and to novel articles of manufacture used in and in combination with such reactors. In neutronic reactors a neutron fissionable isotope such as U^{233}, U^{235}, or 94^{239} or mixtures thereof is subjected to fission by absorption of neutrons, and a self-sustaining chain reaction is established by the neutrons evolved by the fission. In general such reactors comprise bodies of compositions containing such fissionable material, for example, natural uranium, disposed in a neutron slowing material which slows the neutrons to thermal energies. Such a slowing material is termed a neutron moderator. Carbon, beryllium, and D_2O (heavy water) are typical moderators suitable for such use. Heat is evolved during the reaction which is removed by passage of a coolant through the reactor or in heat exchange relationship therewith. Specific details of the theory and essential characteristics of such reactors are set forth in copending application of Enrico Fermi and Leo Szilard, Serial No. 568,904, filed December 19, 1944, now Patent No. 2,708,656.

This invention is particularly concerned with neutron bombardment of various target isotopes. As the result of such bombardment, many substances absorb neutrons to form various useful isotopes. For example, radioactive atoms may be prepared by fission or neutron absorption which may be used as tracers in chemical and biochemical or biological work. Moreover, fissionable isotopes such as U^{233} may be produced by neutron bombardment of an isotope such as Th^{232}.

While a neutronic reactor is an excellent source of neutrons, the problem of using neutrons developed in a reactor for bombardment of target isotopes frequently is rather difficult. Insertion of neutron absorbing target isotopes directly into a neutronic reactor for bombardment purposes offers certain difficulties since the neutron absorption of the inserted isotope may be so great that the chain reaction is stopped unless but a limited amount of such isotope is inserted.

In accordance with the present invention, it has been found that a target isotope or target isotopes may be bombarded with neutrons from a neutronic reactor by disposing the isotope to be irradiated around the exterior of the reactor in association with a neutron moderating or scattering material. In such a process the problem of stopping the reaction is minimized and neutrons which might otherwise be lost are used in the bombardment.

The target isotope or compositions containing such isotope may be disposed as a more or less uniform suspension or dispersion in the neutron moderator. Thus a solution or slurry of the target isotope in a liquid moderator may be used for this purpose. Alternately lumps or granules of the target isotope or a composition containing this isotope may be dispersed in a moderator which may be solid or liquid.

Thus an object of this invention is to provide a novel means for bombarding substances with neutrons. Other objects and advantages of the invention will become ap-

2

parent by reference to the following disclosure and the accompanying drawings in which

Fig. 1 is a flow diagram showing the invention as embodied in a system wherein the reactive composition as well as the bombarded composition is in fluid form;

Fig. 2 is a diagrammatic plan view of a system with the upper portion of the shield broken away wherein the reactive composition and the bombarded composition are solid in form; and

Fig. 3 is a sectional view taken on the line 3—3 of Fig. 2.

In order that a self-sustaining neutronic chain reaction can be established and maintained, the losses of neutrons must be held to a value that at least one neutron is available for a new fission, after losses have been deducted, per neutron consumed in production of fission. In fission of U^{235} and similar isotopes, more neutrons are evolved per fission than are required to produce the fission. For example, about 2.3 neutrons are evolved per neutron consumed in fission of U^{235}, and about 2.8 neutrons are evolved per neutron consumed in fission of 94^{239}. These evolved neutrons are used up in fission of further U^{235} or 94^{239} atoms or are lost. If losses do not reduce the ratio of neutrons evolved to neutrons consumed or lost below one, the chain reaction will continue.

Losses may be external, as when neutrons escape from the reactor, or internal. Internal losses are caused by absorption of neutrons by atoms which do not fission when the neutron has been absorbed.

U^{238} present in natural uranium absorbs substantial quantities of neutrons to produce 94^{239}. This loss may be substantially reduced by use of uranium aggregates. Thus, it has been found that U^{238} absorbs neutrons to an appreciable degree at energies (resonance energies) greater than thermal energies due to its relatively high capture cross-section with respect to that of U^{235} at such resonance energies. However, this type of absorption, known as resonance absorption, may be reduced by decreasing the amount of neutrons which pass into a uranium body until these neutrons have been slowed to thermal energy. This may be done by reducing the ratio of surface area per unit weight of uranium, i.e., by using natural uranium in the form of aggregates preferably having a minimum thickness of about 0.5 cm. Moreover, this loss may be rendered negligible by use of a concentrate of a fissionable isotope which contains greater than natural concentration of fissionable material.

Neutron moderators also absorb neutrons. Generally speaking, it is desirable to use as a moderator an element (or compound thereof) of low atomic weight and low neutron capture cross-section. The ability to slow down neutrons may be expressed by what is known as the scattering cross-section whereas the ability to absorb or capture neutrons may be expressed as the capture cross-section. The ratio of absorption cross-section to scattering cross-section of various materials are approximately as follows:

Light water (H_2O)	0.00478
Diphenyl	0.00453
Beryllium	0.00127
Graphite	0.000726
Heavy water (D_2O)	0.00017

For natural uranium it is preferred to use materials wherein the above ratio is below about 0.004. However, with enriched uranium compositions containing more than natural amounts of U^{235}, a greater latitude is permissible. Using carbon or deuterium oxide as moderators and natural uranium as the fissionable composition, only about 1.1 or 1.3, respectively, neutrons are obtained per neutron consumed due to neutron losses in the U^{238} and the

2,986,510

3

moderator. Since the external neutron losses may be substantial, other internal neutron losses should be held sufficiently low to prevent these losses from rising so high as to prevent the reaction.

Other components of the reactor including the coolant, impurities in the uranium or other portions of the system, moderator, control or limiting rods, fission fragments, restraining barrier, etc. absorb neutrons in varying amounts depending upon their neutron capture cross-section.

The effect of these impurities or absorbers in a reactor containing natural uranium as the fissionable component has been approximately evaluated for each element as a danger coefficient. This coefficient is computed according to the formula

$$\frac{\sigma_i}{\sigma_u} \frac{A_u}{A_i}$$

where

σ_i represents the cross-section for absorption of thermal neutrons of the impurity;

σ_u represents the cross-section for absorption of thermal neutrons of the uranium;

A_i represents the atomic weight of the impurity of neutron absorber; and

A_u represents the atomic weight of uranium.

The following table gives presently known values for various elements having their natural isotopic content.

Element	Danger Coefficient	Element	Danger Coefficient
H¹	10	Mo	1.0
D²	0.01	Ru	~2
He	0	Rh	50
Li	310	Pd	~2
Be	0.04	Ag	18
B	2,150	Cd	870
C	0.012	In	54.2
N	4.0	Sn	0.18
O	0.002	Sb	1.6
F	0.02	Te	1
Ne	<3	I	1.6
Na	0.65	Xe	<6
Mg	0.48	Cs	8.7
Al	0.30	Ba	0.30
Si	0.26	La	<2.4
P	0.3	Ce	<2.4
S	0.46	Pr	<2.4
Cl	31	Nd	~17
A	~0.8	Sm	~1430
K	2.1	Eu	435
Ca	0.37	Gd	~6320
Sc	<7	Tb	~20
Ti	3.8	Dy	~200
V	4	Ho	~10
Cr	2	Er	~40
Mn	7.5	Tm	~20
Fe	1.5	Yb	~10
Co	17	Lu	~30
Ni	3	Hf	~20
Cu	1.8	Ta	4.6
Zn	0.61	W	2.7
Ga	~1	Re	~18
Ge	(<5)	Os	<1.7
As	2	Ir	~70
Se	6.3	Pt	~2.5
Br	2.5	Au	16
Kr	<6	Hg	82
Rb	~0.4	Tl	0.5
Sr	0.4	Pb	0.03
Y	0.57	Bi	0.0025
Zr	~0.13	Th	1.1
Cb	<0.4		

From the above it will be apparent that certain elements such as cadmium, boron and gadolinium absorb neutrons to a very high degree and if present in substantial amount will stop the reaction. On the other hand, larger amounts of other materials may be present. In any case, however, only a limited amount of impurities and target isotopes may be present within the reactor in order to secure a chain reaction.

From the above it will be apparent that for a neutron chain reaction to remain self-sustaining the equation

$$n-x-y-z-L \geq 1$$

4

where

n = number of neutrons evolved by a fission of a fissionable isotope per neutron consumed by such isotope.

x = number of neutrons absorbed by a non-fissioning isotope such as U^{238} in formation of a fissionable isotope per neutron consumed in fission.

y = number of neutrons absorbed by the moderator per neutron consumed in fission.

z = number of neutrons absorbed by other neutron absorbers per neutrons consumed in fission.

L = number of neutrons lost by leakage per neutron consumed in fission.

Thus, with U^{235} the sum of $x+y+z+L$ cannot exceed about 1.3 and with 94^{239} cannot exceed about 1.8.

The ratio of the fast neutrons produced in one generation by the fissions to the original number of fast neutrons producing the fission in a system of infinite size from which there can be no loss is called the reproduction factor and is denoted by the symbol k. The k constant of a system of finite size is the reproduction factor which the system would have if expanded to infinite size. Usually this constant is expressed without regard to localized neutron absorbers such as control or limiting rods, which are not uniformly dispersed throughout the entire system. The neutron reproduction ratio (r) is an actual value for a finite system, and differs from k by a factor due to loss of neutrons through leakage and through absorption by localized neutron absorbers. To maintain a chain reaction, r must be at least equal to one. As pointed out in the above-mentioned Fermi-Szilard application, it is preferably maintained below about 1.01 during operation of the reactor.

Computation of k for any system may be determined experimentally in accordance with methods described in co-pending application of E. Fermi, Serial No. 534,129 filed May 4, 1944, entitled "Nuclear Chain Reacting System," now Patent 2,780,595, dated February 5, 1957.

The reproduction ratio (r) may be ascertained by observation of the rate of increase of neutron density. It may also be predicted by computation of losses due to local absorbers or leakage which may be deducted from k to secure this value. In such a case allowance for leakage is made depending upon the size of the reactors. For reactors of practical size, leakage usually amounts to about 0.01 to 0.3 k units depending upon the amount by which the k of the system exceeds one. Loss due to other absorbers may be computed by computation of the danger sum as heretofore described.

The reactor diagrammatically illustrated in Fig. 1 comprises a suspension such as a solution or slurry of a fissionable material in a liquid moderator. For example, a solution of uranyl sulphate in water is chain reacting where the solution fills a spherical reactor 12 inches in diameter and surrounded with an efficient neutron reflector and where the reactor contains at least 575–600 grams of U^{235} as a uranium concentrate containing about 15% U^{235} based upon the total uranium. Higher amounts of U^{235} for example 600–700 or more grams are required where the reflector contains a target isotope as in the present instance. Moreover, other solutions such as solutions of uranyl fluoride, uranyl nitrate, plutonyl sulphate (PuO_2SO_4) may be used where the fissionable isotope content of the plutonium or uranium is above about 5–15% of the uranium or equivalent composition. In addition, natural uranium compounds (uranyl fluoride UO_2 or U_3O_8) may be dispersed or dissolved in heavy water (D_2O) to establish a chain reaction.

The shape of the reactor may be cylindrical as shown by the drawings or may be spherical or other form. A slender elongated cylinder is particularly advantageous where maximum leakage is desired in order to secure a maximum of neutrons for bombardment of the target isotope or isotopes.

In Fig. 1 a reactor tank or chamber 2 of low neutron

2,986,510

5

absorbing material, for example stainless steel or aluminum, is provided. This tank is sufficiently thin (for example ⅛ inch or less) to permit passages of neutrons therethrough without substantial absorption of neutrons. The tank contains a chain reacting liquid composition 4 such as above mentioned.

The reactive composition is continuously circulated through a heat exchanger 6 by means of a pump 8 having its suction side connected to the tank 2. The discharge side of the pump is connected to the heat exchanger 6 through which the reactive composition is passed in heat exchange relationship with a coolant circulated through the heat exchanger by inlet and outlet pipes 10 and 12. The cooled composition is returned to the tank 2 through a return line 14.

The amount of reactive composition within the tank 2, as well as the concentration of uranium-containing material in the composition, is controlled by a system including a reversible delivery pump 16 connected to the bottom of the tank 2 and to a reservoir 18 having an inlet 20 to accommodate the introduction of uranium-containing material into said reservoir 18. The reservoir 18 is connected to the tank 2 through a line 22 having a conventional three-way operating valve 24 connected to the discharge side of a pump 26, the suction side of which is connected to a moderator reservoir 30 having an inlet 32 through which moderator may be conveyed to this reservoir.

The reactive composition 4 is continuously withdrawn from the bottom of the tank 2 through an outlet line 34 connected to the suction side of a pump 36, the discharge side of which is connected to a conventional separator device 38 adapted to separate the moderator from the uranium-containing material. Such a separator may comprise an evaporator or settling tank and the separated material may be conveyed from the device 38 by an outlet line 40 for recovery of 94 and fission products formed as a result of the neutronic reaction where natural uranium or U²³⁵-U²³⁸ mixtures are used within the tank 2. Separated moderator is conveyed from the separator device 38 through a line 42 including a pump 44, the discharge side of which is connected to a moderator purifier 46 from which the purified heavy water is conveyed to the before-mentioned reservoir 30. The purification may be effected by various means such as by distillation.

It will be understood that the water within the tank 2 is continuously decomposed into D_2 and O_2 or H_2 and O_2, depending upon the type of water used, as a result of the neutronic reaction; and these decomposition products, as well as gaseous fission products of the reaction, are swept from the tank 2 and the decomposition products are recombined. A gas pump or blower 48 is provided having its suction side connected to a helium reservoir 50, and its discharge side connected to the tank 2 above the level of the reactive composition therein. The helium passes through the tank 2 and is conveyed therefrom by an outlet line 52 connected to a conventional recombiner device 54 adapted to recombine the hydrogen isotope or isotopes and O_2 into vaporized D_2O or H_2O which is conveyed to a condenser 56, the condensed D_2O or H_2O being conveyed to the before-mentioned purifier tank 46. Helium is conveyed from the condenser 56 by a line 57, preferably including a pump or blower 58, to a helium purifier tank 60 for removal of radioactive impurities and thence to the helium reservoir 50.

An emergency dump line 62 is connected to the tank 2 and the reservoir 18, said line including a dump valve 64 adapted to be opened under emergency conditions to reduce the body of composition 4 within the tank 2 to a size smaller than that at which a chain reaction may be sustained.

It may be noted that the system, thus far disclosed, is purely illustrative and such systems are more fully described in copending application, Serial No. 613,356, filed August 29, 1945 in the United States Patent Office by

6

Eugene P. Wigner, Leo A. Ohlinger, Gale J. Young and Harcourt C. Vernon and also in an application of Robert F. Christy, Serial No. 623,363 filed October 19, 1945, now Patent 2,843,543, dated July 15, 1958.

Surrounding the tank 2 is another tank or chamber 66 within which is a production area or zone containing a fluid composition 68 to be bombarded by neutrons emanating from the reactor 2. This tank also is constructed of a low neutron absorber such as aluminum or stainless steel or other material having a danger coefficient below 10 and having a thickness, e.g. ⅛ inch or less, sufficiently low to prevent substantial neutron absorption by the tank 66. The composition 68 may be a slurry or solution of the neutron absorbent material in a neutron moderator such as heavy water, said composition being admitted to the chamber 66 through an inlet line 70 including a conventional shut-off valve 72. After the composition 68 has been bombarded for the desired length of time, it is conveyed from the chamber 66 by an outlet line 74 including a conventional drain valve 76 to a conventional separator device 78 adapted to separate the heavy water from the neutron absorbent material which has at this point been converted to a radioactive isotope by the capture of neutrons as above discussed. The radioactive material is conveyed from the separator 78 by a line 80, and the separated moderator is conveyed from the device 78 by a line 82 connected to the suction side of the before-mentioned pump 44 which thus urges the heavy water from the line 82 to the before-mentioned purifier tank 46. It will be understood that the presently illustrated system will be used when the same moderator is used in tanks 2 and 66 and that where different moderators are used different purifiers may be required.

The bombarded composition 68 is preferably circulated through a heat exchanger 84 by a pump 86 having its suction side connected to the chamber 66, the cooled composition being returned to the chamber 66 through a return pipe or line 88.

The chambers 2 and 66 are disposed within an aluminum or steel tank or chamber 90 containing a heavy water neutron reflector 92 adapted to reflect escaping neutrons back into the chamber 66 and/or reactor 2. The heavy water 92 is conveyed to the chamber 90 through an inlet line 94 having a conventional shut-off valve 96 and is conveyed from the tank by an outlet line 98 having a conventional drain valve 100.

Thus, it will be understood that by the above described system a novel method and means have been provided for surrounding a neutronic reactor with a fluid composition to be bombarded by neutrons escaping from the periphery of the reactor, said composition including a neutron scattering or reflecting material combined with a neutron absorbent material adapted to be converted to a radioactive isotope by absorption of neutrons.

Referring now to Figs. 2 and 3, the neutronic reactor diagrammatically shown therein comprises a body of neutron moderator 102, preferably in the form of graphite or beryllium oxide blocks, in which are disposed a plurality of spaced lumps or slugs 104 (Fig. 3), of uranium-containing material, the portion of the graphite outwardly of the slugs constituting a reflector 106 (Fig. 3) for reflecting a substantial number of the escaping neutrons back into the central portion of the reactor. A control rod 108 of highly neutron absorbent material, such as cadmium or boron, extends through a complementary slot in the reactor to accommodate control of the neutron density therein as more fully brought out in the Fermi-Szilard application above-mentioned.

The reactor is contained within a tank or chamber 110 which, in turn, is disposed within a tank or chamber 112, both of these tanks being constructed of neutron permeable material such as aluminum and being contained within a concrete vault 114. The space between the tanks 110 and 112 is filled with a plurality of blocks 116, 116 composed of material to be bombarded by neutrons

2,986,510

7

emanating from the reactor, each of these blocks being provided with a handle **118** to facilitate insertion and removal thereof. The blocks **116, 116** are preferably formed of graphite or other solid moderator material mixed with a neutron absorbent material to be converted to a radioactive isotope by absorption of neutrons.

Within the vault **114** around the tank **112** is a reflector **120** preferably formed of blocks or graphite, said reflector functioning to reflect escaping neutrons back into the blocks **116, 116** thereby increasing the rate of neutron absorption by the blocks. Thus, it will be understood that in the embodiment illustrated in Figs. 2 and 3, the neutronic reactor is surrounded by inner and outer reflectors **106** and **120** with a layer of material **116** of bombarded material interposed between the inner and outer reflectors. The inner reflector **106** functions to reduce neutron leakage from the exterior of the reactor; and thus, the reactor is capable of sustaining a chain reaction even though it is somewhat smaller than the critical size at which such a reaction would normally be possible without the use of such a reflector. The reflector **120** serves to reflect escaping neutrons back into the blocks **116, 116**, thereby increasing the rate of neutron absorption by the bombarded material.

It is, of course, obvious that numerous variations are available without departure from the scope of the invention. For example, the air or water cooled graphite moderated neutronic reactor described in the aforementioned Fermi-Szilard application may be used to generate neutrons for bombardment as herein contemplated. A typical reactor of this type comprises a cube of graphite provided with holes extending horizontally therethrough provided with means to circulate water through the holes and having metallic uranium bodies therein. The holes are lined with aluminum pipe and the uranium jacketed with an aluminum sheath.

The principal dimensions of the reactor are as follows:

Axial length of active cylinder of reactor=7 meters
Radius of active cylinder of reactor=4.94 meters
Total weight of uranium metal in rods=200 metric tons
Weight of graphite in reactor=850 metric tons
Radius of uranium metal rods=1.7 centimeters
Thickness of aluminum jackets=0.5 millimeter
Thickness of aluminum pipe=1.5 millimeters
Thickness of liquid layer=2.2 millimeters with water or 4 millimeters of diphenyl
Number of rods in reactor=1695
Weight of aluminum in reactor=8.7 metric tons
Rod spacing in square array=21.3 centimeters

As a further modification, a neutronic reactor containing aggregates of uranium and moderated with deuterium oxide may be constructed using a tank of aluminum 6 feet in diameter and 7 feet 4 inches high. In one such reactor 136 rods of uranium metal 1.1 inches in diameter and having an aluminum jacket 0.035 inch thick are mounted vertically in the tank to extend to within ¼ inch from the bottom of the tank. The ractor is surrounded with a 12 inch reflector of graphite. When 122.4 centimeters of D_2O containing less than 1 percent H_2O is placed in the tank, the reactor reaches critical size. When 124.7 centimeters of D_2O is introduced, the time for doubling of the neutron density therein is about 6.5 seconds. A dispersion of the target material and moderator may be disposed about this type of reactor in place of the graphite reflector or in conjunction therewith and bombardment of the target isotope thereby obtained.

While the bombardment preferably is conducted using a target material disposed in a neutron slowing material, this is unnecessary particularly where a neutron reflecting

8

layer is interposed between the active portion of the reactor and the material bombarded. Thus, since the graphite section **106** serves as a neutron reflector, the blocks **116** may consist substantially entirely of the material to be subjected to bombardment.

Various materials may be bombarded. For example, graphite bodies may be bombarded to increase their electrical resistance, thermal conductivity and elastic modulus or to remove impurities such as boron. Various isotopes may be bombarded to form other isotopes. For example, thorium 232 may be bombarded to form U^{233}. Other elements such as sulphur, phosphorus, boron, cadium or compounds thereof may be irradiated as will be understood by the art. Deuterium oxide containing one or more percent of light water may be bombarded to convert the light water to heavy water at least to a substantial degree. Where a neutron moderator is used in combination with the target material the moderator preferably should have a neutron absorption cross section less than that of the target material.

While the theory of the nuclear chain fission mechanism in uranium set forth herein is based on the best presently nown experimental evidence, the invention is not limited thereto, inasmuch as additional experimental data later discovered may modify the theory disclosed. Any such modification of theory, however, will in no way affect the results to be obtained in the practice of the invention hereindescribed and claimed.

Obviously, many modifications may be made in the specific embodiments disclosed without departing from the intended scope of the invention.

What is claimed is:

A massive leakage irradiator comprising a central core of thermal neutron fissionable material and moderating material, and a reflector free of fissionable material, the said central core in the shape of a solid cylinder and the reflector composed of a first zone and a second zone, said first zone being in the form of a first hollow cylinder with open ends immediately and coaxially surrounding the solid cylinder of the core along the entire length of its curved side and composed of cylinder segments with handles embedded in their upper surface, and the second zone being in the form of a second hollow cylinder with open ends of the same height as the first hollow cylinder and the solid cylinder of the core immediately and coaxially surrounding the first hollow cylinder along the entire length of its curved side, the first hollow cylinder containing reflecting material intermixed with material to be irradiated, and the second hollow cylinder being composed only of reflecting material, whereby the massive leakage of neutrons through the exterior curved side of the cylinder of the core may be utilized for irradiation.

References Cited in the file of this patent

UNITED STATES PATENTS

2,708,656   Fermi _____ May 17, 1955

FOREIGN PATENTS

861,390   France _____ Oct. 28, 1940
114,150   Australia _____ May 2, 1940
233,011   Switzerland _____ Oct. 2, 1944
648,293   Great Britain _____ Jan. 3, 1951

OTHER REFERENCES

Kelly et al.: Physical Review 73, 1135–9 (1948). Copy in Patent Office Library. (204/154.2).

A General Account of the Development of Methods of Using Atomic Energy (1940–1945). H. D. Smyth. For sale by Supt. of Documents, Washington, D.C. Pages 153, 177.

A 1991 Package of Declassified Information on the "Water Boiler" Reactor

Los Alamos
Los Alamos National Laboratory
Los Alamos, New Mexico 87545

DATE: February 18, 1991
IN REPLY REFER TO: HSE-6-WB
MAIL STOP: F691
TELEPHONE: 665-2814

I am pleased to send you a packet of information related to the Water Boiler Event at Los Alamos on November 5, 1990. At this event the American Nuclear Society recognized the Los Alamos Water Boiler Reactor as a Nuclear Historic Landmark. The enclosed packet is a collection of various items generated in connection with the event that the steering committee thought you might find enjoyable. A list of the enclosures along with some explanatory comments prefaces the packet.

Based on feedback to committee members, we feel that the event was very successful. Please contact me if I can be of further assistance.

Sincerely,

Norman L. Pruvost
Chairman, Water Boiler
Steering Committee

LIST OF ENCLOSURES

1. Copy of WATER BOILER Plaque

2. Copy of WATER BOILER Citation

3. Brief History — The Water Boiler Reactor 1944–1974, by Paul W. Henriksen, written for Water Boiler Ceremony.

4. Graph depicting estimates of point of criticality

5. EARLY REACTORS — written by M. Bunker published in LOS ALAMOS SCIENCE, 1983 – Fortieth Anniversary Edition.

6. Sections on Water Boiler from PROJECT Y: The Los Alamos Story, by David Hawkins. (Tomash Publishers, Los Angeles, Ca., 1983).

7. Excerpt on Water Boiler by Paul W. Henriksen from Hoddeson, Henriksen, Westfall, and Meade, "Critical Assembly," LANL document to be published.

8. Copies of pictures and articles excerpted from newspapers and Los Alamos Newsbulletin.

9. Address List of Water Boiler Event attendees or interested colleagues.

10. Copy of LANL Museum Log, November 5, 1990.

THE LOS ALAMOS WATER BOILER REACTOR 1944 – 1974

The Water Boiler Reactor was conceived and built during the Manhattan Project to obtain needed critical mass data. The reactor evolved from a critical assembly that produced on May 9, 1944, the first self-sustaining nuclear chain reaction using enriched uranium.

Designated as a Nuclear Historic Landmark, May 1990 by the American Nuclear Society

AMERICAN NUCLEAR SOCIETY
555 North Kensington Avenue, La Grange Park, Illinois 60525 USA
Telephone: (708) 352-6611   •   Telecopier: (708) 352-0499   •   Telex: 4972673

JOSEPH C. BRAUN Ph.D.
Executive Director

November 5, 1990

Dr. James F. Jackson
Deputy Director
Los Alamos National Laboratory
P.O. Box 1663, MS A101
Los Alamos, New Mexico   87545

Dear Dr. Jackson:

In May 1990, The American Nuclear Society's Board of Directors approved a Nuclear Historic Landmark plaque for The Los Alamos Water Boiler Reactor and authorized presentation of the bronze plaque inscribed:

THE LOS ALAMOS WATER BOILER REACTOR

1944-1974

The Water Boiler Reactor was conceived and built during the Manhattan Project to obtain needed critical mass data. The reactor evolved from a critical assembly that produced on May 9, 1944, the first self-sustaining nuclear chain reaction using enriched uranium.

As a further explanation of the Water Boiler Reactor's place in history, the Board approved this citation:

The reactor, code named "Water Boiler" for security reasons, was a hollow stainless steel sphere, one foot in diameter. It contained a solution of 14.5% enriched uranium sulfate and water and was surrounded by a reflector of beryllium oxide. A control rod, safety rod, and neutron detectors completed the assembly. It was to be the world's third nuclear reactor[1], the first homogeneous liquid-fuel reactor, and the first reactor to be fueled by uranium enriched in the isotope ^{235}U.

Following its first criticality on May 9, 1944, minor refinements were made on the assembly, and additional measurements were made. By the end of June, the original assembly had fulfilled its goals; it had provided a way for experimentally checking theoretical methods for calculating critical mass, it provided the means for measuring how various reflector materials affected the critical mass, and it provided experience in assembling a supercritical system.

---

[1]The first two were Fermi's "pile" at Chicago's Stagg Field and the X-10 graphite reactor at Oak Ridge.

2. Dr. Jackson   November 5, 1990

Because of its lack of shielding and the absence of a heat-removal system, the original assembly was not designed to operate at any appreciable power level. It could not provide a sufficiently strong source of neutrons needed by the Laboratory for nuclear data measurements and other studies. A higher power version of the Water Boiler was constructed soon after. This version operated at a power level of 5.5 kilowatts. The sulfate solution of the original assembly was replaced with a nitrate solution and cooling coils were placed inside the spherical core. In addition, a 'glory hole' through the core allowed samples to be placed in the intense neutron field with the core. Massive concrete shielding was added to surround the reactor. This version of the Water Boiler began operation in December, 1944.

In March of 1951 the third, and final, Water Boiler version went into operation. This reactor operated at 30 kilowatts and provided the higher neutron fluxes desired by the research staff. It contained additional cooling coils within the 1-foot diameter core for greater heat-removing capacity. The enrichment of the uranyl nitrate solution was increased from 14.5% to 88.7%. Also, the original beryllium oxide reflector was replaced with graphite. In addition, a hydrogen-oxygen recombination system was connected to the fuel vessel. This system eliminated the hazard posed by the evolution of hydrogen and oxygen from radiolytic breakdown of the water during operation. This final version of the Water Boiler operated almost daily until its deactivation in 1974.

The plaque is presented today, November 5, 1990, in Los Alamos, New Mexico, at a special presentation ceremony hosted by the Trinity Local Section. The persons and organizations who contributed to the success of The Los Alamos Water Boiler Reactor are to be congratulated.

Sincerely,

Joseph C. Braun

Joseph C. Braun

JCB/jh

xc: R. D. O'Dell
        G. M. Montoya
        N. L. Pruvost
        N. M. Trahey

The Water Boiler Reactor 1944-1974

One of the primary goals of the Manhattan Project was to enrich uranium with the isotope ^{235}U. While this material was destined for the core of an atomic bomb to be built at Los Alamos, the first tiny shipments from Oak Ridge were not simply stockpiled, but were used as the fuel for a simple self-sustained, nuclear, chain-reacting assembly. Aptly named the "Water Boiler" to disguise its true purpose, it was an experimental test of theoretical calculations of chain-reacting assemblies of uranium; however, subsequent higher power versions of the Water Boiler were true research reactors. The Water Boiler was the first critical assembly to be built from enriched uranium and the third distinct nuclear chain reacting system overall. It worked well from the start, and its three versions continued to provide valuable scientific information for 30 years.

The Water Boiler was first envisioned in April 1943 and championed by Robert Bacher as a means of generating a critical mass at the earliest possible moment. Others argued that it wasn't needed because such a device was not close enough in design to metallic weapon cores, but Bacher reminded them that critical assemblies were so new that any test of their construction and calculation would be valuable. Indeed much could be learned in such a new area of science and technology just from the construction and operation of the reactor. The Governing Board decided to build a prototype reactor to operate at a nominal power of 1 W so that cooling and shielding would not be needed. Donald Kerst was given the task of designing the reactor, but there were few people at Los Alamos with reactor experience. Nevertheless within a year the reactor was operating.

The first version of the Water Boiler, known as "Lopo" (for low power), was designed to use the minimum amount of enriched uranium; thus the container was made spherical and the uranium was put in a water solution. The other materials used in its construction absorbed a minimum of neutrons and a good neutron reflector was placed around the sphere. The simplicity of the stainless steel sphere was important in minimizing the critical mass. The uranium sulfate "soup" was contained in a conical pan and was fed into the sphere through a vertical pipe. The reactivity of the solution was increased in increments by removing a small amount of the solution (beginning with distilled water) and adding more concentrated solution to the conical pan under the sphere and mixing it. The new soup was then forced up by air pressure to fill the sphere and enter a pipe connected to the top of the sphere. The reactor went critical on May 9, 1944 with 565.5 g of a 14% ^{235}U solution at 39°C. Enrico Fermi was at the controls and several other famous physicists and chemists were present that day. The final amount of enriched uranium in the critical mass was very close to the predictions of the theorists.

Once Lopo had verified the critical mass calculations and given Los Alamos practice in constructing a critical assembly, Fermi and Bacher pushed for the construction of a higher power version to serve as a source of neutrons that would roughly represent the neutron spectrum from a weapon. Such a reactor could produce strong neutron fluxes at a moderate power level. Lopo was therefore dismantled and a heavily shielded, 5-kW version (nicknamed Hypo) was designed and built. The sphere thickness was doubled, a cooling coil was added inside the sphere, a tube was placed

1

through the sphere to give access to the highest flux region, and the fuel solution was changed from sulfate to nitrate. The beryllium oxide bricks were needed for other experiments, so some of them were replaced by graphite blocks. The completed reactor provided a copious source of neutrons, but it did not operate at such a high power that contamination from the fission gasses and cooling were serious problems. The neutrons were used, for example, to make needed cross-section measurements on various elements.

In 1950, the progression to higher operating powers continued with the conversion of Hypo to Supo. The conversion involved more modifications, but retained the basic design. One of the changes, which was a testimonial to the usefulness of the reactor, was an increase in the number of research facilities, which involved enlarging the building to give access to more faces of the reactor. Other major design changes included the addition of a hydrogen-oxygen recombiner to remove an explosive hazard and the replacement of the remaining beryllium oxide reflector with graphite. A second round of design changes involved improvements in the safety and operation of the facility. This version of the Water Boiler was the longest lived, operating almost continuously until 1974. The reactor has since been dismantled, and the building that housed it has been decontaminated.

The impact of the Water Boiler was not confined to Los Alamos. Spinoffs from it included a 1-W solution-type reactor constructed by and for North American Aviation (NAA) and a 500-watt reactor constructed by NAA for Lawrence Livermore Laboratory. The Water Boiler was also the prototype for the first reactor not to be owned and operated by a national government: a 10-kW research reactor for North Carolina State College.

Self-Sustaining Nuclear Chain Reactions

	Unit Name	Reacting System	Location	Date of First Criticality
1.	Chicago Pile 1	Natural uranium, graphite	Univ. of Chicago	Dec. 2, 1942
	Chicago Pile 2	Natural uranium, graphite	Argonne, IL	March 1943
2.	X-10 Reactor	Natural uranium, graphite	Oak Ridge, TN	Nov. 4, 1943
3.	**Water Boiler**	**Enriched uranium, water**	**Los Alamos, NM**	**May 9, 1944**
4.	Chicago Pile 3	Uranium, heavy water	Argonne, IL	May 15, 1944
5.	B-Pile	Nat. uranium, graphite (water cooled)	Hanford, WA	Sept. 27, 1944

While the Water Boiler is listed here as the third reactor, it was the first reactor to deviate from the original design of Fermi, et al., at Chicago: a heterogeneous uranium graphite pile. Chicago Pile 2 was simply Chicago Pile 1 moved to the Argonne Laboratory site. The X-10 reactor at Oak Ridge involved many design changes over Chicago Pile 1 as it was scaled up to production size.

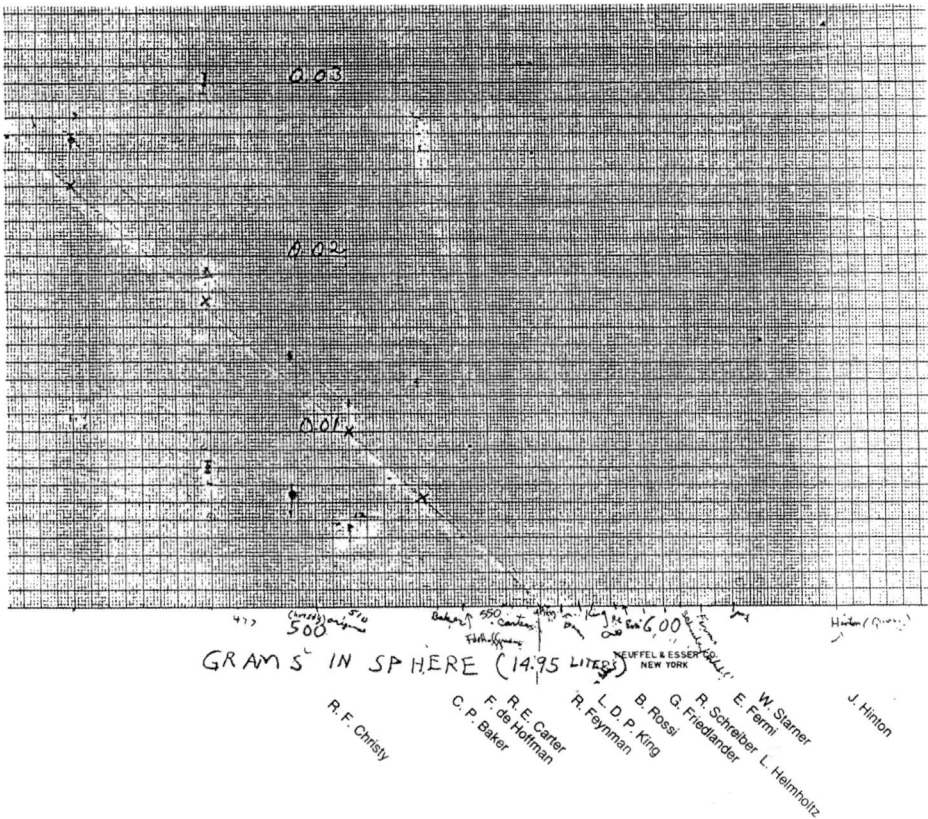

GRAMS IN SPHERE (14.95 LITERS)

R. F. Christy
C. P. Baker
F. de Hoffman
R. E. Carter
R. Feynman
L. D. P. King
B. Rossi
G. Friedlander
R. Schreiber
E. Fermi
L. Helmholtz
W. Starner
J. Hinton

The above graph was included without explanation with the water boiler reactor description. It appears to show measurements of how close the reactor was to becoming critical (the unlabeled vertical axis) as the amount of uranium in the reactor was increased (the horizontal axis, showing the mass of uranium in grams). The line fitted through the X symbols intercepts the horizontal axis at roughly 565 grams of uranium, indicating that this should be the critical mass of uranium for this reactor. A fainter line with a slightly different slope, presumably from a different set of measurements, has almost the same intercept. It was 565.5 grams of uranium that actually yielded criticality on May 9, 1944. (Robert Christy's initial calculations had indicates that the critical mass would be 600 grams. When he obtained improved neutron cross sections to use as input to his calculations, his prediction of the critical mass became 575 grams, within 2% of the actual value.)

Project Y: The Los Alamos Story

PART I

Toward Trinity

BY DAVID HAWKINS
With a New Introduction by the Author

PART II

Beyond Trinity

BY EDITH C. TRUSLOW
AND RALPH CARLISLE SMITH

Tomash Publishers
LOS ANGELES / SAN FRANCISCO

CHAPTER 6

The Experimental Physics Division

March 1943 to August 1944

ORGANIZATION

The Experimental Physics Division was one of the first organized. These were the initial groups.

P-1	Cyclotron Group	R. R. Wilson
P-2	Electrostatic Generator Group	J. H. Williams
P-3	D-D Source Group	J. H. Manley
P-4	Electronics Group	D. K. Froman
P-5	Radioactivity Group	E. Segre

Two new groups under H. Staub and B. Rossi were created in July and August 1943. The first was to develop improved counters; the second, improved electronic techniques. In September they were combined as the Detector Group, P-6, under Rossi. Group P-7, the Water Boiler Group, under D. W. Kerst, was created in August. R. F. Bacher was Division Leader from the time of his arrival in July 1943.

EQUIPMENT

When the first members of the experimental physics groups arrived in March 1943, the buildings to house the accelerating equipment were not completed.

The bottom piece of the Harvard cyclotron was laid on April 14, and in the first week of June there were initial indications of a beam. The early work, with an internal beam on a beryllium target probe, gave an

THE WATER BOILER

The first chain-reacting unit built at Los Alamos was the Water Boiler, a low-power pile fueled by uranium enriched in ^{235}U. It was the first pile built with enriched material, the so-called alpha-stage material containing about 14% ^{235}U. The necessary slowing or moderation of fission neutrons was provided by the hydrogen in ordinary water. The active mixture was a solution of uranyl sulfate in water. The tamper was beryllia.

The Water Boiler was to provide a strong neutron source of experiments and to serve as a trial run in the art of designing, building, and operating such units. It was an integral experiment to test a theory similar, in some respects, to that involved in designing a bomb. It was the first of a series of steps from the slow reaction first produced in the Chicago pile to the fast reaction in a sphere of active metal. It laid the foundation for instrumental and manipulatory techniques required in later, more exacting steps. Unfortunately, Los Alamos workers did not have the full benefit of experience gained by those at Chicago, so there was unnecessary delay before the first chain reaction was started.

Water Boiler calculations absorbed much of the Theoretical Division's time. Calculation of the critical mass depended upon applying diffusion theory to a complex system of active solution, container, and tamper. To conserve material, it was important to find the optimum solution concentration. The number of hydrogen nuclei had to be large enough to slow the neutrons to thermal energies and small enough not to capture too many of them.

The Water Boiler was isolated for safety reasons. It was first planned for 10-kW operation. The radioactivity of fission fragments from intermittent operation was estimated at 3000 curies (Ci). The minimum safe distance from unprotected people was calculated on the assumption that a mild explosion could disperse this activity into the atmosphere. Isolation was also desirable because of possible high instantaneous radiation in case of an uncontrolled chain reaction.

In September 1943, while design of the Boiler and the building to house it were still preliminary, Fermi and Allison came from Chicago to discuss the problems of such a unit. They pointed out many difficulties in operating the Boiler as a high-power neutron source. Some had been anticipated but their acuteness had not been appreciated fully. One problem was gas evolution that would cause unsteady operation. Decomposition of the uranium salt and consequent precipitation would result from the large amount of radiation to which it would be subjected. Heavier shielding than had been planned would be necessary.

These discussions led to omitting all features necessary for high-power operation and going ahead with the design of a low-power Boiler. Provisions were made, however, for later installation of equipment for high-power operation. The main omission was equipment for chemical decontamination, unnecessary for operation at trivial power outputs. The Boiler could no longer be used as an intense neutron source, but it could be used to investigate a chain-reacting system with a much higher ^{235}U enrichment than previous piles.

The building to house the Boiler, associated laboratories, and later critical assemblies in Los Alamos Canyon at Omega Site (Fig. 3) was completed in February 1944. Design problems included a heavy concrete wall to separate the Boiler from remote-control equipment; a thermostated enclosure to maintain constant Boiler temperature; recording and monitoring equipment, including ionization chambers and amplifiers; control rods and their associated mechanisms; a support for the tamper and container, the container itself, together with means for putting in and removing solution; and design of beryllia bricks for ease in fabricating and stacking the tamper. Specifications for the tamper and active solution were worked out with the Radiochemistry and Powder Metallurgy Groups (Chap. 8), after the original choices of material and size were made by the theorists.

Tests of the fluid-handling and counting equipment late in April 1944 ended with use of normal uranyl sulfate solution. Enough enriched material had arrived to permit determining the critical mass, after the chemists purified and prepared the solutions. Successful operation of the

Water Boiler as a divergent chain reactor was a small but important step toward controlled use of nuclear energy from separated ^{235}U or plutonium.

The Water Boiler, like other controlled reactors, depended upon the very small percentage of delayed neutrons. These allowed keeping the system below critical for prompt neutrons and near critical for all, including the delayed neutrons. Although the delayed neutrons are only about 1% of the total, in the region near critical the system's time dependence (rate of rise or fall) is only about the duration of the delay period. Prompt chains die out constantly, to be reinstated only because of the delayed neutrons.

A Water Boiler experiment proposed by Rossi and bearing his name was designed to determine the prompt period. This period depends on the time it takes for the neutrons to be emitted after fission, on the fission spectrum, and on the scattering and absorption characteristics of core and tamper. It was essential to measure the prompt period in a metal assembly as accurately as possible. Its measurement in a hydrogenous assembly would not give direct information relevant to efficiency calculation, but would provide experience and instrumental development and also would be a check on theoretical predictions.

The Rossi experiment counts neutron coincidences. The presence of a prompt chain in the reactor is presumed whenever a neutron is counted. A time-analyzing system then records the number of neutrons counted in short intervals immediately after the first count. This gives a direct measure of the prompt period.

Another method that gave less interpretable results was to change the degree of criticality rapidly by a motor-driven cadmium control vane. A third experiment was to measure the spatial distribution of neutrons in the solution and tamper by placing small counters in various positions in the Boiler and tamper. This experiment served as a check of calculation from neutron diffusion theory.

A fourth, Rossi-related experiment was to measure fluctuations in the Boiler neutron level. These measurements were of interest relative to the variation of the neutron number from fission to fission, which was, in turn, related to the statistical aspects of the chain reaction in the bomb, particularly the predetonation probability. The first measurement gave the count fluctuation in a counter relative to the average number of counts. This gave information about the neutron number fluctuation as soon as the effective number of delayed neutrons was measured.

About mid-1944, the Water Boiler Group planned to make critical assemblies with uranium hydride and to rebuild the Water Boiler for higher power operation.

107

The list of enclosures for the information package indicates that the following description of the water boiler reactor is an excerpt (containing the account by Paul W. Henriksen) from the Los Alamos National Laboratories document "Critical Assembly" by Hoddeson, Henriksen, Westfall, and Meade:

C. The Water Boiler – The Beginning of Critical Mass Studies

A reactor using ^{235}U in a water solution, the "Water Boiler," had been proposed in April 1943, as part of a program to measure critical masses from the neutron multiplication of chain reacting systems. Bacher argued strongly that the Laboratory should "start with something that would have a smaller critical mass, like a water solution," which could be built sooner than a ^{235}U metal assembly.

---

[63] Williams, "Cross Sections for Fission of 25, 49, 28, 11, 00, 02, B and Li," LA-150, 5 Oct. 1944, pp. 6, 9.
[64] Williams, "Cross Sections for Fission of 25, 49, 28, 11, 00, 02, B and Li," LA-150, 5 Oct. 1944, pp. 10, 11.
[65] Inglis, "Experiments Related to the Fission Process," and Williams, "Competition Between Capture and Fission," p. 138 in "Nuclear Physics," LA-1009, pp. 123, 131.

234

UNCLASSIFIED

The Water Boiler could not indicate the critical mass of a weapon with a ^{235}U metal core, since critical mass depends on the chemical, physical, and geometrical conditions. Nonetheless, it could check the theory for calculating critical masses, determine the effect of various tamper materials on critical mass, and give people experience in assembling a supercritical system. "I had really quite a battle with some of the theorists on this," Bacher explains, since they thought critical mass studies with chain reacting systems unlike that to be used in the weapon were a "waste of time." But Bacher prevailed, and plans for the Water Boiler proceeded in summer 1943. This early step in critical mass studies would be followed, as material became available, in 1944 and 1945, by neutron multiplication measurements of uranium hydride cubes, by the construction of a higher power reactor, and ^{235}U and ^{239}Pu metal assemblies (chapters 14 and 17).[66]

In early summer 1943, Kerst, who was appointed to head the Water Boiler project, sketched out a design for the world's first chain reacting system using enriched material. The reactor was a simple hollow stainless steel sphere about one foot in diameter filled with a solution of a sulphate of ^{235}U and water, surrounded by a BeO reflector, and neutron shielding. No one doubted that it would "go critical" when enough enriched uranium from Oak Ridge was put into it. The main question was the amount of enriched ^{235}U that would make it critical.[67]

Kerst, a University of Illinois physicist known for his invention of the betatron, had no direct experience with building a chain reacting pile and could draw on little expert help. Fermi was able to give advice only intermittently, and as Kerst recently explained, the only people with experience in critical assemblies were at Oak Ridge and the University of Chicago, and they were not available to Los Alamos on a full time basis in 1943. Los Alamos had to educate its own critical assembly experts, and the Water Boiler provided the experience. The Water Boiler crew came primarily from Purdue where they had been working on measurements for the Super. As Raemer Schreiber, one early crew member, recalls: "There weren't any experts in nuclear reactors, so they were picked up from cyclotron people and nuclear physicists."[68] The key crew members involved in assembling and operating the low power water boiler were: Kerst (group leader), Baker, Gerhart Friedlander, Lindsay Helmholz, Marshall G. Holloway, L. D. P. King, and Schreiber.[69] Help with the general theory came from Christy, whose calculation of the Water Boiler's critical mass became known to the Water Boiler staff as the "Bible."[70]

The location of the Boiler was an important consideration; it needed to be

[66] Bacher interview by Westfall, 14 Dec. 1987, OH-169, p.7. M. Holloway interview by Henriksen and Hoddeson, 9 Sept. 1985, OH-60; Christy interview by Hoddeson, 14 April 1986, OH-117; and Frisch, Hanson, Anderson, "Los Alamos Technical Series Volume 5: Critical Assemblies, Part 1," LA-1033, 19 Dec. 1947, p. 1-2; R. Schreiber interview by Hoddeson, 16 Aug. 1984, OH-48.
[67] Hawkins, "Manhattan District History, Project Y, the Los Alamos Project," *LAMS-2532*, p. 116.
[68] R. Schreiber interview by Hoddeson, 16 Aug. 1984, OH-48.
[69] "An Enriched Homogeneous Nuclear Reactor," *Review of Scientific Instruments*, 22 (July 1951), p. 489; Holloway interview by Henriksen and Hoddeson, 9 Sept. 1985, OH-60; Kerst interviews by Hoddeson, 8 Aug. 1985, OH-54, and 28 April 1986, OH-118.
[70] Holloway interview by Henriksen and Hoddeson, 9 Sept. 1985, OH-60; Schreiber interview by Hoddeson, 16 Aug. 1984, OH-48.

accessible from the Technical Area, but distant enough so that radiation leaks or other disasters would not contaminate the town.[71] Oppenheimer favored a site in upper Los Alamos Canyon just down the cliff from the town and downstream from the water supply. Christy calculated the probable area that radioactive materials would cover if an accidental explosion occurred, and Los Alamos Canyon seemed to provide that margin of safety. The Governing Board approved the site at the 19 August 1943 meeting.[72]

The most important technical decision in the summer of 1943 was the operating power level. As originally conceived, the Water Boiler would have operated at ten kilowatts, powerful enough to provide a strong source of neutrons.[73] Although a strong neutron source would have been useful for some experiments, objections to high power operation surfaced quickly in the Governing Board. The following arguments were made: protecting operators from the high radiation would be difficult; the ^{235}U would be contaminated with radioactive fragments which would need to be chemically removed; a complicated cooling system would be needed (since the enriched uranium was so scarce, one could not have large reservoirs of it outside the reactor; all of it had to be used at once); and the uranium "soup" was sure to bubble, making the neutron output fluctuate even more unpredictably than normal.[74]

A quiet debate ensued throughout the month of September in the Governing Board. Proponents of high power operation stressed the importance of a powerful source of neutrons, while Bacher and others worried that —since the high power operation would raise the radioactivity level to several thousand curies — the active material would have to be decontaminated to allow experiments at a lower radioactivity level, such as with the hydride. The power proponents won out temporarily as tentative plans were made to operate at 10 kw, but the push for high power operation collapsed a few weeks later at the 30 September meeting. Samuel K. Allison and Fermi pointed out difficulties in operating at power that did not bear on the ultimate goal of the laboratory, and mentioned that the Met Lab no longer needed a high neutron flux to study poisoning effects.[75]

Christy calculated that 600 grams of pure ^{235}U would be critical in an infinite water tamper.[76] The Water Boiler group took this prediction and plunged ahead to make a more detailed design. The essential elements of the Water Boiler were mechanically simple. Schreiber and King developed a stainless steel fluid handling system to form the heart of the reactor. The "soup" rested in a conical stainless steel reservoir whose shape was not optimum for producing a chain reaction, and which allowed access to the solution so that its concentration could

---

[71] The Trinity test would require a similar, but scaled up site search in 1945.

[72] Governing Board Minutes, 22 July 1943 and 19 Aug. 1943, A-83-0013, 1-18 and 1-22.

[73] P-Division Progress Report, 15 Aug. 1943, LAMS-7.

[74] Governing Board Minutes, 5 Aug. 1943, A-83-0013, 1-20.

[75] Governing Board Minutes, 9 and 30 Sept. 1943, A-83-0013, 1-24 and 1-27. The slow pace of the ^{235}U production would probably not have allowed the high power operation, but that was not evident until early in 1944, well after this decision was made. Schreiber interview by Hoddeson, 16 Aug. 1984, OH-48, and King interview by Hoddeson, 16 Aug. 1984, OH-49.

[76] Christy, "Critical Mass of ^{235}U in Water Solution with Water Reflector," LAMS-18, 4 Oct. 1943.

be increased as the ^{235}U arrived at Los Alamos. The soup was pumped by air pressure into a sphere surrounded by a tamper where the reaction took place. The entire system was closed to the atmosphere so that any radioactive gas generated during the reaction would not be accidentally expelled. A control rod, similar to, but smaller than the control rods on the Chicago pile, extended into the tamper shell and could be quickly dropped in to quell a runaway reaction.[77]

An important chemical decision had to be made concerning the type of uranium salt to use in the Boiler. Chemist L. Helmholz experimented with various compounds, but the choice finally came down to nitrate and sulfate.[78] The choice was important since the enriched radioactive salt had to be completely soluble within the stainless steel sphere. In mid November, the sulfate was chosen, since it had the desirable features of being more soluble, and absorbing fewer neutrons than the nitrate.[79]

The Water Boiler, though simple in theory, was difficult to translate into stainless steel and beryllium oxide. The stainless steel sphere was difficult to make, since the hemispheres could not be joined by soldering; the solder would have been corroded by the acid soup. The company making the sphere for Los Alamos did not fully understand that requirement and soldered it for lack of a substitute method. Los Alamos ultimately arc welded the sphere.[80] The size of the sphere was also important; if it was too small, a critical mass could not be put in it and if too large, the geometry might not allow a critical mass at the earliest possible time. The final selection of a 12-inch diameter sphere required that a new contour be ground into the BeO tamper blocks (by mounting the polar segment on the faceplate of a 36-inch lathe and grinding it with a diamond wheel).[81]

The tamper that surrounded the sphere, to reflect neutrons back into the soup, was in fact made from pure beryllium oxide powder pressed into molds to form dense bricks. This material oxide was the original choice for the tamper in the bomb, but it was soon learned that it could not be made dense enough to fit in the small bomb and still reflect the neutrons properly.[82] Pure BeO proved to be difficult to procure and fabricate into bricks, in part due to backlogs in the shop.[83] The neutron detectors, the most complex part of the system, required effort from everyone in the group. The Water Boiler team had time to solve these problems because the ^{235}U was delayed in arriving. Although the Laboratory thought ^{235}U shipments would be well along by late 1943, in November the Governing Board did not anticipate one kg of 10% ^{235}U being on hand until January 15.[84] Actually, sufficient ^{235}U did not arrive until April 1944.[85]

---

[77] King, "Design and Description of Water Boiler Reactors," paper presented at the International Conference on the Peaceful Uses of Atomic Energy, 30 June 1955.

[78] Dodson to Kennedy, CM-Division Progress Report, 1 Oct. 1943, LAMD-71.

[79] P-Division Progress Report, 1 Nov. 1943, LAMS-25; CM-Division Progress Report, 15 Nov. 1943, LAMD-71.

[80] King interview by Hoddeson, 16 Aug. 1984, OH-49; P-Division Progress Report, 15 Jan. 1944, LAMS-50.

[81] Schreiber interview by Hoddeson, 16 Aug. 1984, OH-48.

[82] Schreiber interview by Hoddeson, 16 Aug. 1984, OH-48.

[83] King interview by Hoddeson, 16 Aug. 1984, OH-49.

[84] Governing Board Minutes, 11 Nov. and 1 Dec. 1943, A-83-0013, 1-33 and 1-36.

[85] Schreiber interview by Hoddeson, 16 Aug. 1984, OH-48. One novel problem consisted of accurate temperature

The Omega building became usable, even though incomplete, on 1 February 1944, and the reactor materials came together in March; the BeO bricks arrived acceptably pure, the stainless steel spheres were properly welded, and the fluid handling equipment installed.[86] By 1 April the sphere had been fitted into the tamper. Five different neutron counting devices were placed in different positions with respect to the central sphere of the boiler.[87] The Water Boiler crew then began fine tuning the equipment by running experiments with unenriched uranium and a radium-beryllium source. By 1 May enough enriched material was on hand to start sub-critical experiments.[88] The Water Boiler, with its reactivity controlled by cadmium rods, went critical on 9 May 1944, first with a neutron source in the center and later in the day without the source.[89]

The Water Boiler group then worked for several weeks to refine their apparatus, while Christy refined his prediction of the critical mass.[90] With new data on cross sections, Christy made a very quick and dirty correction to his original calculation of 600 grams, and came up with an estimate of approximately 575 grams. The group removed tubes in the tamper and filled holes used to insert a ^{238}U chamber into the center, thereby making the tamper shell more uniform. They found the critical mass of the uranium sulfate solution in the Boiler to be 565 grams.[91] As King later remarked, the agreement between theory and experiment was reached "probably somewhat fortuitously." Nonetheless, this success helped raise confidence in T-Division's ability to calculate critical mass (section A).[92] Other studies in May measured the period of the reactor as a function of control rod setting and measured the effect of tamper materials on the critical mass.[93]

By June 1944, the Water Boiler had fulfilled its goals: it had provided a way to check theoretical methods for calculating critical mass, the means for measuring how tamper materials affected critical mass, and experience in assembling a supercritical system. It was time to focus attention on new projects. By this time, the group had already started work on hydride critical assemblies (chapter 14). Another major project was the high power Water Boiler, which would be constructed, in part, from dismantled equipment from the first Water Boiler. The

---

control for the Water Boiler room. Changes in temperature as small as 0.001 C° would produce noticeable changes in the reactivity of the solution. The Water Boiler was enclosed in a special room maintained at a constant temperature with an electronic thermostat control designed by Matthew Sands of the Electronics group. Temperature changes in the boiler were held to within 0.01°C. King, "Los Alamos Technical Series V. 5: Critical Assemblies, Part 2," LA-1034, 19 Dec. 1947, p. IV-7; P-Division Progress Report, 1 May 1944, LAMS-95, p. 2; Oppenheimer to Bacher, 27 Jan. 1944, A-84-019, 14-7; CM-Division Progress Report, 1 May 1944, LAMS-86.

[86] P-Division Progress Report, 1 Feb. 1944, LAMS-53.

[87] P-Division Progress Report, 15 May 1944, LAMS-96; and Baker, H. K. Daghlian, Friedlander, "Water Boiler," LA-134, 8 Sept. 1944, pp. 22-25.

[88] P-Division Progress Reports for 1 March, 15 March, 1 April, and 1 May 1944, LAMS-58, 75, 78, and 95.

[89] Hawkins, "Manhattan District History, Project Y, the Los Alamos Project," *LAMS-2532*, p. 118; King, "Critical Assemblies, part 2," V. 5, chapter 4.

[90] P-Division Progress Report, 15 May 1944, LAMS-96.

[91] P-Division Progress Report, 1 June 1944, LAMS-111.

[92] King, "Critical Assemblies, Part 2," LA-1034, V. 5, chapter 4, p. IV-17; Christy interview by Hoddeson, 14 April 1986, OH-117, p. 10.

[93] P-Division Progress Report, 1 June 1944, LAMS-111.

Festivities for historic Water Boiler Reactor draw crowd of Manhattan Project pioneers

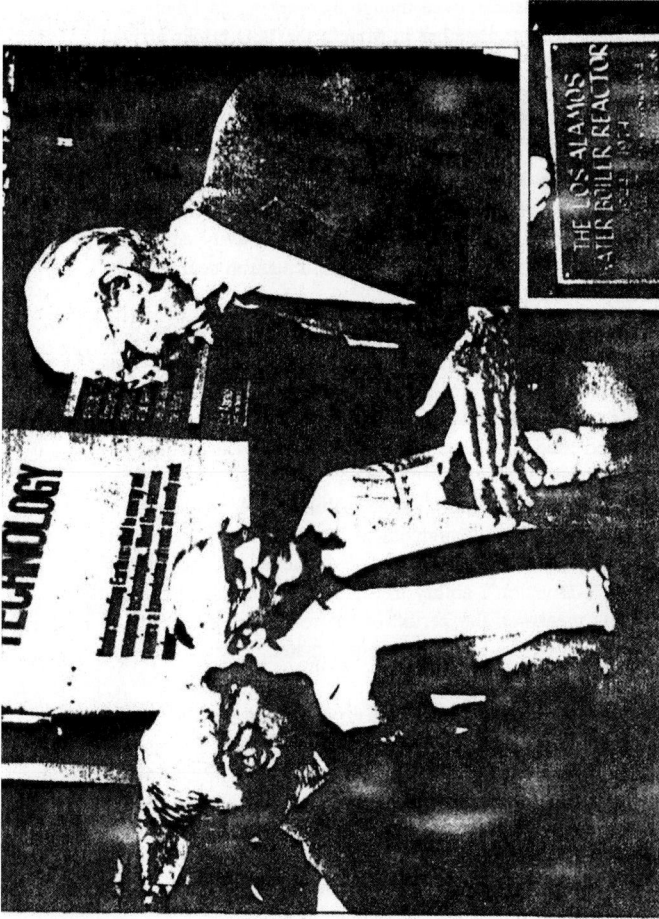

L.D.P. "Perc" King (left) reminisces with Harold Hammel (center) and Donald Kerst at a special ceremony held this week in recognition of the Los Alamos Water Boiler Reactor. The reactor, which has been decommissioned, recently was named a nuclear historic landmark by the American Nuclear Society. (See the Oct. 26 Newsbulletin for more details.) The ceremony at the Bradbury Science Museum attracted close to 100 individuals who worked on three separate versions of the reactor between 1943 and 1974. About a third of the people in attendance came from out of state to be reunited with friends and former co-workers, some of whom they hadn't seen in decades. Kerst, who recently retired from the University of Wisconsin and moved to Vero Beach, Fla., was leader of the group that designed and built the original Water Boiler criti-

cal assembly in 1943 and 1944. Hammel, a resident of Ellettsville, Ind., was a technician who worked with Kerst on the prototype. King, a Tesuque resident who also was a member of the original group, eventually became leader of the group that designed and built the second and third versions of the reactor. The final version served as a strong neutron source for the Laboratory's research staff until 1974. The inset photo shows the commemorative plaque that was placed at a site in Los Alamos Canyon where the reactor once stood. Photo by Fred Rick

Los Alamos
Los Alamos National Laboratory
Los Alamos, New Mexico 87545
(505) 667-7000

public information group

news release

CONTACT: John A. Webster

MEDIA ADVISORY

LOS ALAMOS, N.M., Nov. 2, 1990 -- Los Alamos National Laboratory is having a public ceremony in recognition of its decommissioned Water Boiler Reactor, recently named a nuclear historic landmark by the American Nuclear Society.

The reactor was built during the Manhattan Project and operated until 1974. It was code-named Water Boiler for national security reasons to conceal its intended purpose and also because of its design.

A public ceremony to recognize the reactor's new status is scheduled from 10:30 to 11:30 on Monday, Nov. 5, in the Bradbury Science Museum on Diamond Drive.

Several of the Manhattan pioneers who worked on the original reactor are expected to attend. Jack Ohanian, president of the American Nuclear Society, will present a commemorative plaque and citation to James F. Jackson, the Laboratory's deputy director. There also will be a special display of historic photographs and other reactor memorabilia dating back to the 1940s.

Reporters and photographers are welcome. If you plan to attend this event, please contact Kathy Haq at 667-7000 in the Laboratory's Public Information Office.

Los Alamos National Laboratory is a multidisciplinary research facility which applies science and technology to problems of national security ranging from defense to energy research. It is operated by the University of California for the Department of Energy.

-30-

ANNOUNCEMENTS FOR UPDATE/INFORM
to be used Monday-Wednesday-Friday-Monday (Oct. 29 and 31, Nov. 2 and 5)

Friday, Oct. 26
Promote contents of Newsbulletin, including story on the Water Boiler Reactor.

Monday, Oct. 29
The Los Alamos Water Boiler Reactor, conceived and built during the Manhattan Project, has been designated a nuclear historic landmark by the American Nuclear Society. See last Friday's Newsbulletin for details.

Wednesday, Oct. 31
A public ceremony to commemorate the landmark status of the Los Alamos Water Boiler Reactor is planned for 10:30 a.m. this coming Monday at the Bradbury Science Museum. See last Friday's Newsbulletin for details.

Friday, Nov. 2
Join retired physicist Perc King and Manhattan pioneer Donald Kerst Monday at the Bradbury Science Museum to help celebrate the American Nuclear Society's newest historic landmark: the Los Alamos Water Boiler Reactor. Kerst, a University of Illinois physicist who was selected to oversee the critical assembly in 1943, and King, who oversaw the reactor, will be joined by others who helped make Enrico Fermi's concept a reality. A public reception is scheduled from 10:30 to 11:30 a.m. at the museum. See the reminder in today's Newsbulletin and the Oct. 26 issue for details.

Monday, Nov. 5
Drop by the Bradbury Science Museum between 10:30 and 11:30 a.m. today to meet the pioneering scientists who invented and operated one of the world's earliest reactors. The Los Alamos Water Boiler Reactor has been designated a nuclear historic landmark by the American Nuclear Society. The society's president, Jack Ohanian, will be on hand to present a plaque and commemorative citation to Deputy Director Jim Jackson.

A Retrospective Article in the Pasadena Star-News

The following article of Sunday August 6, 1996 is reproduced courtesy of the Pasadena Star-News:

Star News

TRAVEL

A string of villages along California's coast offer funky fun.
D10

VOICES

SUNDAY, AUGUST 6, 1996

D1

INSIDE

NY Times Best Sellers	D2
Bookings	D3
Joal Ryan	D7
Dave Barry	D7
Daily Horoscope	D8
Ann Landers	D8
NY Times Crossword	D8
Travel Notes	D10

HIROSHIMA

Helping change course of history

By Tom Scanlon
STAFF WRITER

What immortal hand or eye
Could frame thy fearful symmetry?
. . . What the hand dare seize the fire?
William Blake, "The Tyger"

ROBERT Christy's long hands, his hazel eyes and his keen mind helped frame the most fearful man-made devastation of our world. Manhattan Project scientists like Christy — after whom the prototype plutonium bomb was named — dared seize the fire of atomic fission, developing two atomic bombs dropped in a week of terror that began 50 years ago today. In 72 hours, Hiroshima and Nagasaki were in ruins, more than 100,000 citizens were dead, and just as many were injured.

Christy, featured in the Showtime production "Hiroshima" (a mixture of dramatization and interviews airing on the cable channel today at 8 p.m.), has no regrets about the history of which he was a part.

"I'm comfortable with the work we did. I feel we did make a significant contribution to ending the war," Christy says, speaking in a low, confident voice in his quiet Pasadena home near Caltech. After 40 years as a professor there, he retired in 1986.

"I'm happy that I was able to make some contribution to that."

Here, the eloquent 79-year-old scientist pauses, carefully measuring his words, as if judging two test tubes. The price of the Manhattan Project, he says, was "major loss of life in Japan. But I firmly believe the price to Japan would have been worse if we hadn't finished the war with the bomb."

In any case, it was not in Christy's power — nor in that of any of the Manhattan Project scientists — to

Staff photo by LEO JARZOMB

ROBERT CHRISTY, of Pasadena, was among the scientists who worked on the Manhattan Project.

Please turn to CHRISTY / D5

CHRISTY

Continued from D1

decide whether to use the bomb. The decision as to
how, when and where to use what Christy and the
others created belonged to President Harry Truman
and a select group of military leaders.

Important as he was in snapshots of Los Alamos,
in the big picture Robert F. Christy was but a foot
soldier on the frontline of the nuclear arms race.

A 26-year-old Christy didn't really know much
about the Manhattan Project he was asked
to join. "As far as I was concerned, anything sounded more interesting than what I was
doing . . . I believe I was told not long afterward we
were trying to design nuclear reactors in order to
make plutonium."

A native of Canada, he had received his doctorate
degree from UC Berkeley, and was teaching at the
Illinois Institute of Technology when he was recruited to the Metallurgical Laboratory. In 1942, Christy
accepted an offer to join the new Manhattan Engineering District of the U.S. Army, so named because
the first head of the project had offices in New York
City.

"We were all anxious to do our part in the war for
two reasons," Christy recalls, from a breakfast nook
overlooking his back yard. "One, we believed in the
war effort. And the other (reason) is if we weren't
doing that, we would've been drafted into the Army.
I think most of us preferred to be working on the
Manhattan Project than be drafted and be in the
trenches."

In the winter of 1943, J. Robert Oppenheimer —
tapped by Gen. Leslie Groves to supervise the scientific aspects of the top-secret project — went to
Chicago on a recruiting mission. He asked Christy,
a student of his at Berkeley, to join noted scientists
from around the world who were starting a new
division of the Manhattan Project in New Mexico.

Christy was flattered, and didn't think twice
about taking Dagmar, his new wife, to Los Alamos, a
secluded mesa 40 miles from Santa Fe. Shortly after
arriving there, Christy learned of the goal of this
great experiment: creation of atomic bombs.

IN stark contrast to what it would bring to the world, Los Alamos was a peaceful, secluded place in the mid-1940s. "I enjoyed the countryside there very much," Christy recalls.

Oppenheimer, a mentor and something of a father figure for the orphaned Christy (his father died when he was 2, his mother when he was 10), encouraged his horseback riding by giving him half of one of his horses. Christy shared the horse with another scientist, but "we never could figure out who owned the head and who owned the rear end."

Christy also took up skiing and often went on long hikes with Hans Bethe, the well-known physicist who left Nazi Germany in the 1930s.

At first, the Los Alamos scientists believed they were in a race with Germany, fearful that Hitler would use the power of nuclear bombs to put the world under his heel.

One of Christy's first tasks was to try to find the critical mass of uranium 235 that would be needed in a "water boiler" to create a nuclear chain reaction. According to "Critical Assembly: A Technical History of Los Alamos," the work by the Canadian scientist was indispensable:

"Help with general theory came from Robert Christy, whose calculation of the Water Boiler's critical mass came to be known as 'The Bible' among the Water Boiler staff."

On May 9, 1944, Christy's calculations proved quite accurate in testing, as the Los Alamos water boiler "went critical." This was only the third nuclear reaction ever on record, and the first to use enriched uranium.

By now, Germany was all but defeated, but war raged on in the Pacific, and the American top-secret effort continued, with the belief that "the gadget" would end the war.

The Los Alamos scientists were confident they could make a successful uranium bomb, triggered by the "gun method" — firing a bullet containing uranium into a uranium target. More tricky was the design of a bomb involving plutonium, which can be produced in chain reactions (unlike uranium, which is scarce). But early experiments showed the "gun method" was not fast enough to work with plutonium.

In July of 1943, Los Alamos scientist Seth Neddermeyer proposed an "implosion method" that would use a hollow sphere of plutonium (or uranium) surrounded by explosives which, when detonated, would cause the plutonium to implode in on itself, setting in motion the atomic chain reaction.

The problem in the implosion method was in try ing to get the sphere to collapse symmetrically, believed to be necessary for fission to occur.

A few months after the water boiler reaction, Christy was to make another important contribu- tion to the Manhattan Project, when he proposed a radically different design for an implosion bomb. According to "Critical Assembly," ". . . in late Sep- tember theoretician Robert Christy suggested using a solid rather than hollow core . . . (Italian scientist Bruno) Rossi was immediately enthusiastic about Christy's suggestion.

"The brute force of Christy's design simply cir- cumvented the serious symmetry and stability prob- lems of a more elegant implosion design."

This is confirmed by another book, "Project Y: The Los Alamos Story" (the second volume of "The History of Modern Physics 1800-1950"):

"Near the end of 1944, the solid sphere or Christy implosion was selected."

Thus, the implosion bomb came to be known as the "Christy bomb" or "Christy gadget," much to the disgust of Neddermeyer. "At Los Alamos they always called it the Christy bomb," he is quoted as saying, in "Lawrence and Oppenheimer," a book written by Nuel Pharr Davis. "The *Christy* bomb!"

IN July of 1945, nerves were tight around Los Alamos, as preparations began for a test of the Christy gadget. After some delays, the test was conducted at a site in remote Alamogordo that Oppenheimer dubbed Trinity.

The pre-dawn test on July 16, 1945 was a wild suc- cess, and Christy will never forget the sight of the world's first atomic bomb detonating. "It was awe inspiring." Twenty miles from the bomb's center, he witnessed a flash of light "brighter than the sun."

Similar explosions of light proved deadly at Hiroshima and Nagasaki — surprising even those scientists who created the bombs.

Before Hiroshima, "We went into some discus- sion about the scenarios of damage," says Christy. "We were partly misled, because most bombs involve damage to buildings, knocking down build- ings. Most of our calculations had to do with the dis- tance of buildings damaged . . .

"But the flash of light from the bombs turned out to be more injurious than the effects of the blast . . .

people out in the open were badly burned. My recollections are that we had not calculated that effect. One of the most upsetting things were pictures of how badly people were burned."

On Aug. 9, with fires still burning from the Christy bomb that exploded over Nagasaki, Japan surrendered to the United States.

With the war ended, Christy returned to Chicago for a few months at the University of Chicago. In 1946, Oppenheimer used his influence to bring Christy to Caltech as a theoretician. While Oppenheimer would leave the following year, Christy remained at the university for 40 years, serving as academic vice president from 1970-80.

Over the years, he would be joined at Caltech by several Los Alamos alumni: Richard Feynman (now deceased), Robert Bacher, Robert Walker, Matthew Sands and Dave Wood.

In 1960, Christy took a sabbatical, as Oppenheimer gave him a research fellowship at Princeton University. There, Christy turned his attention to astrophysics, and "realized techniques we developed at Los Alamos to calculate implosion could be used to calculate the behavior of variable stars."

CHRISTY feels his work in astrophysics is much more scientifically refined than anything he did during World War II. While the importance of his work is underscored in detailed books on the Manhattan Project, Christy is not overly proud of his Los Alamos achievements.

"I was not irreplaceable — no one was irreplaceable there." His work on the atomic bomb, he adds, "was engineering, more than discovery."

But what engineering it was. At the starting line of the nuclear arms race, the then-young Christy joined a superstar-filled relay team that brought atomic fusion theory into horrifying reality.

Fifty years ago, Christy was a young, unproven scientist working under great pressure; no one asked him what he thought of the bomb, no one cared. Now, after turning the events over in his mind for a half-century, Christy's cold, scientific conclusion is that we were right to use the bombs.

"I had no reason to disagree, and I still don't disagree. I think that, under the circumstances, we did the right thing."

He points to estimates that a U.S. invasion of Japan might have cost as many as one million U.S. soldiers — a controversial figure that some now consider inflated for propaganda purposes. One thing that is indisputable is the toll of U.S. "fire bombing" of Japan. The March 9, 1945, fire bombing of Tokyo, for instance, killed some 80,000 civilians.

If the atomic bombs brought war to an end quickly, then they may have actually saved lives — in the short run, at least.

Five years ago, Christy visited Hiroshima and Nagasaki, as part of his work with the Radiation Effects Research Foundation, a joint U.S.-Japanese study of the effect of the atomic bombs dropped on Japan. There are some scientific effects of atomic bombs that "still defy us to some extent," Christy says.

THESE days, the lanky Christy — 6-foot-4, 175 pounds — plays tennis, rides horses and generally leads an active life with astrophysicist Juliana Sackmann, his German-born second wife. They have two daughters, Juliana and Alexandra, both attending Stanford and both horse enthusiasts. Christy was divorced from his first wife (now deceased), with whom he had two sons, Thomas, a physician, and Peter, a computer designer.

Pondering the future, Christy is optimistic that his children and his two grandchildren will not witness the escalation of the nuclear destruction he helped create.

Current atomic weapons, he notes, "are up to 1,000 times more powerful than what we set off — we're lucky they have not been used. The major powers, I think, now are convinced a war with atomic bombs is not feasible. You'll get destroyed yourself."

His personal optimism, however, is cut with a grim scientific view.

"We can destroy most of the (atomic) bombs we have . . . but still the danger is there. Once you know how to build atomic bombs, it is not fantastically difficult to do it."

What the hammer? what the chain?
In what furnace was thy brain?
What the anvil? what dread grasp
Dare its deadly terrors clasp?

William Blake, "The Tyger"

Chapter 7

Opposing Nuclear Weapons Proliferation

Robert's View of the "Christy gadget"

The initial impetus for the U.S. atomic bomb project had been the fear that the Nazis in Germany were working to design nuclear weapons for use in the war. However, Germany lost the war in May of 1945, before the U.S. bombs were completed. In the meantime the war in the Pacific had become very bitter. As described in the previous chapter, there had already been huge American and Japanese military losses, as well as huge Japanese civilian losses from conventional bombing.

Robert Christy (Caltech photo)

The Hiroshima atomic bombing on August 6, 1945 (using the "gun-type" uranium-235 bomb) has been estimated to have resulted in between 90,000 and 160,000 deaths, in total. The Americans had hoped that the shock of the atomic bombing would cause the immediate end of the war. However, when the Japanese generals did not surrender within the next three days, the Nagasaki bomb (of the "Christy gadget" type, using plutonium-239) was dropped on August 9, 1945, with an estimated total death toll of about 80,000. This finally brought an end to the Second World War.

It was clear that if the war had continued, then there would have been much larger numbers of deaths, both on the American side and the Japanese side. An invasion of Japan itself was being planned by the Americans, and it was expected that millions of lives would be lost in such an invasion. Robert was at peace with his conscience, knowing that far more lives had been saved than had been lost when the war was terminated by the dropping of the atomic bombs.

However, Robert strongly opposed any further use of nuclear weapons. As described below, he continued this opposition for the rest of his life.

The Association of Los Alamos Scientists

Robert Christy was one of the founding members of the Association of Los Alamos Scientists (ALAS), which was founded on August 30, 1945, only a few weeks after the bombs had been dropped. As described in its manifesto, this organization's purpose was "to promote the attainment and use of scientific and technological advances in the best interests of humanity." The founding scientists believed that "by virtue of their special knowledge, [they had], in certain spheres, special political and social responsibilities beyond their obligations as individual citizens." Their primary concern was to influence public policy in the hopes of promoting international control of atomic energy, with the intent of channeling it into peaceful uses.

Another goal was the public promotion of science. This was pursued both by promoting federal legislation in that area via correspondence with government officials, and via a program of public education on the nature and control of atomic energy using press releases, lectures, films, and exhibits.

The Oppenheimer Trial

Edward Teller was ambitious, and had always wished to be the head of the Manhattan Project and be in charge of building an atomic bomb. He had been displaced from his home country, Hungary, and hated the Communists. Even though he had come to Los Alamos to work on the atomic bomb project there, he stayed largely uninvolved, working on speculative calculations aimed towards producing a hydrogen bomb, which would be even more powerful — although of course nothing came of those calculations at that time.

With the termination of the Second World War, most of the Los Alamos scientists, including Oppenheimer and Robert, no longer wished to continue building bombs. Their new goal was to prevent an arms race with the Soviets. Teller, on the other hand, pushed for the construction of bigger and better bombs: in 1949, he was instrumental in setting up the project to create the hydrogen bomb.

Oppenheimer originally opposed this project as both unwise and unfeasible. There was no reasonable design for a hydrogen bomb at that time, and he felt that the effort would be better put into creating tactical atomic bombs rather than huge weapons that could only be used on civilian targets. However, Teller was successful in his role as leader of the project: in 1951 a workable design for a hydrogen bomb was created. At this point Oppenheimer supported its creation, feeling that the Soviet Union would certainly build such a bomb too now, since it had proved to be feasible. Teller's hydrogen fusion bomb design was first tested in 1952. It proved to be roughly a thousand

Robert Christy (Caltech photo)

times as powerful as the Hiroshima and Nagasaki fission bombs (and also rather more powerful than the typical hydrogen bombs that were later put into production). Teller's success greatly increased his influence in government circles.

The early to mid 1950's have been described as the "McCarthy era." Accusations of Communist associations led to many Americans losing their jobs or suffering the destruction of their careers; some were even imprisoned. Most of this came about through trial verdicts that were later overturned, laws that were later declared unconstitutional, dismissals for reasons later declared illegal or actionable, and the like. Since then, the term "McCarthyism" has often been used to refer to the practice of making accusations of disloyalty or treason without proper regard for evidence, especially in order to restrict dissent or political criticism.

Oppenheimer had been under secret investigation for years. There was a final hearing in April-May 1954, to determine whether his security clearance should be revoked. Robert said,

> I was not called on as witness in any part of it: only a very few people were called as witnesses, or to testify. We all, however, because we (like many people) knew

and admired Oppenheimer, were very concerned about the trial. We thought it was a very improper thing for the government to be doing, and were very disturbed by the whole affair.

Teller testified that he believed that Oppenheimer was loyal to the U.S., but of questionable judgment. This contributed to Oppenheimer losing his clearance (ironically, one day before it was slated to lapse anyway). He also lost his political power as a leader in science, being removed from his leadership of the Atomic Energy Commission. However, he did continue as Director of the Institute for Advanced Study at Princeton. Robert recalled:

> When the report of the trial came out, as a book-length public document, the book was essentially torn into pieces so that it could be circulated more rapidly among the various people in the department who were interested. You'd get one piece, then you'd pass it on to the next person and get another piece of the book, and in this way we read about the details of what went on at the trial. But the result of this was a very distinct polarization in the physics community.
>
> Probably most of the physicists were strong supporters of Robert Oppenheimer, and felt that he was being unjustly attacked by representatives of the government, and were very upset by the whole affair. There were a few members of the physics community who were more worried about the Communist threat, and they (probably sincerely) felt that Oppenheimer was a threat because he had had left-wing inclinations before the war, and they felt that he could not be trusted. As I said, I didn't question their sincerity but I did question their judgment. The result of this was that there was, as I said, a distinct polarization in the community between those who supported Oppenheimer and those who felt that he should not have his clearance reinstated.
>
> Teller was one of those who supported the government and did not trust Oppenheimer, and as a result of that I had a falling out with Teller and refused to have any dealings with him thereafter.

Robert continued to speak up in public for Oppenheimer, and supported him in every way he could. Robert stated in his interview in 2006 with Sara Lippincott:

> My feelings were very strong. I told you earlier that in some sense I viewed Oppenheimer as a god. He was on a pedestal, and I looked up to him. And I was sure that he was not a treasonable person. I knew he had leftist contacts; that was well known to everyone. But I felt that it was just the wrong thing to do, for an honorable physicist to testify against Oppenheimer. It just wasn't right. And I was very upset by it. I still am. I felt, therefore, that it was really improper, it was wrong.

The Teller Handshake Incident

Shortly after the Oppenheimer hearing, there was a meeting at Los Alamos of the scientists who had worked on the atomic bomb. Robert encountered Teller on the porch of Fuller Lodge, where many of the scientists had gathered. Teller, in his ebullient and enthusiastic way, approached Robert with his hand outstretched to shake hands. The other scientists looked on as Robert, a tall imposing figure, stiffened up, put his hands behind his back, and declared that he was no longer willing to shake hands with Teller. It was a very deliberate action on Robert's part — an impulse of course, because Robert had not had time to plan this — and it was recognized by everyone for what it was: that Robert refused to have a direct association with Teller. The two of them were never on friendly terms after that. Robert said, "This was a rather public gesture, and was noted by a number of people, and I believe that hurt Edward [Teller] very much; but it expressed the way I felt — I reacted according to my feelings."

The handshake incident was more poignant in that Robert and his family had shared a house with Teller and his family for about half a year in Chicago in 1946, when they both took up positions at the University of Chicago following their Los Alamos years (as described at the beginning of the following chapter).

Most of the science community supported Oppenheimer, and this handshake incident became a symbol for them; Teller found himself rather isolated, with little respect from many of his scientific colleagues. He stated later in his life that what he did to Oppenheimer was the biggest mistake of his life, that he regretted it and would not do it again.

In his 2006 interview with Sara Lippincott, Robert said:

> I've seen him [Teller] from time to time. Our relationship has remained cool. Since that time, I have disagreed with him in a number of areas: for example, the Strategic Defense Initiative. I have disagreed with him, but I have not argued with him publicly, because Teller operates at a much different level than I do. He's a confidant of Presidents; I'm not. As I say, I merely disagree privately, and that's the way it is.

Security Interview of Robert in 1955, Regarding Communism

A year after the trial where Oppenheimer's security clearance was revoked, Robert was also investigated and interviewed by the Atomic Energy commission. A 30-page transcript of that interview survives in Robert's files.

Among other things, the interviewer asked whether Robert had attended any meetings directly sponsored by the Communist Party while he was a graduate student at Berkeley, or any time thereafter. Robert vaguely recalled attending a meeting at Berkeley where he saw Communist pamphlets, and which thus may have been a Communist-sponsored meeting. Robert was asked whether he had had a "friendly interest" in Russia and in Communism while at Berkeley in 1937 or 1938. He answered that the circles in which he had moved then had indeed had such a friendly interest, which he had shared to some extent at that time. Robert was asked whether he was fully aware at that time of "the tenet of the Communist philosophy advocating force and violence and the overthrow of governments." He had not been aware of any such tenets in the 1930's, and this was of course a philosophy with which he could truthfully state that he did not agree.

Robert was asked about his association with Averill Berman (a presumed Communist) with whom he had once shared a broadcast debate about the H-bomb. Questions were also asked about his fellow Berkeley graduate student Eldred Nelson (who also had been on the Manhattan Project), and a French post-doctoral fellow called Claude Bloch whom Robert had supervised at Caltech. Finally he was questioned about his wife Dagmar and his long-time friend George Volkoff, both of whom could be considered émigrés from pre-communist Russia, although Dagmar's parents were German and had been in exile in Russia for only a few years before fleeing the Communist Revolution (see Chapter 4).

Unlike Oppenheimer, Robert did not lose his security clearance.

The 1963 Enrico Fermi Award for Oppenheimer

On March 1, 1962, four physicists (Freeman Dyson, T. D. Lee, and C. N. Yang of the Institute for Advanced Study at Princeton and A. Pais of Columbia University) nominated Oppenheimer for the Enrico Fermi Award. This award was established in1954 to be awarded yearly "for any especially meritorious contribution to the development, use, or control of atomic energy." They sent a copy of their nomination letter to a number of colleagues in the hopes that they would also write letters to support Oppenheimer's nomination. Robert Christy of course wrote a supporting letter, saying:

> I wish to endorse the nomination of J. R. Oppenheimer for the Enrico Fermi award. My reasons follow.

My first acquaintance with Robert Oppenheimer was in the period 1937-41 when, as a student of his, I came to admire him immensely and started a friendship which has continued. Since then, I have come to appreciate better the dominance of his role in those years in the training of a new generation of theoretical physicists. His was a major influence in the development of theoretical physics in this country.

I next knew him at Los Alamos during 1943-46. The success of that development can in considerable measure be attributed to the outstanding qualities of his leadership. Since that period, he has continued to exert a strong influence both on the development of theoretical physics and, as long as he was able, on the development of atomic energy.

I believe that we can not continue to make awards in this field without recognizing his paramount contributions.

Oppenheimer was awarded the Enrico Fermi Prize in 1963.

Working Towards an Atmospheric Test Ban

Adlai Stevenson, a Democrat and an intellectual, had run for the Presidency in 1952, before the Oppenheimer trial led Robert to take an interest in politics. Stevenson's opponent was Dwight Eisenhower, who had been the Supreme Commander of the Allied Forces in Europe during the Second World War. Stevenson lost, and Eisenhower became President of the U.S. One of the issues on which they differed was atomic bomb testing.

The very first nuclear bomb test, of the "Christy gadget," had taken place at Alamogordo, New Mexico on July 16, 1945. This test was to confirm that the implosion-type nuclear bomb design was feasible, and to test the actual size and effects of the explosion. Fallout was very poorly understood at that time; the first appreciable understanding was not attained until years after the bombings of Hiroshima and Nagasaki.

The U.S. conducted five more nuclear tests in the four years before the Soviet Union tested their first atomic bomb on August 29, 1949. By the 1950's the U.S. had established a dedicated test site in Nevada and another in the Marshall Islands in the Pacific. With Teller's development of the hydrogen bomb and the increased speed of production of nuclear weapons, nuclear testing took off in the 1950's, both of new hydrogen bomb designs and of improved fission designs. After 1950 roughly a dozen U.S. nuclear tests were performed each year. In addition, the Soviet Union tested several nuclear weapons, primarily in Kazakhstan.

Initially most nuclear bomb tests were atmospheric tests, where an atomic bomb is detonated above the ground. This involved several hazards that could have been avoided by carrying out underground tests instead, particularly the danger from fallout.

The worst early example was the test of the first practical hydrogen bomb design, the U.S. "Castle Bravo" test in 1954. Unexpected interactions resulted in the explosion being more than twice as powerful as had been predicted. This meant that there was much more fallout than had been expected. In addition, a change in the weather pattern spread the fallout over areas that had not been evacuated in advance. There were dangerous levels of radioactive contamination in an area over one hundred miles long that included several inhabited islands. There was one death from radiation sickness of a crewman on a Japanese fishing boat, and the islands' inhabitants suffered from radiation burns. Later effects included increased rates of cancer and birth defects among those who had been exposed.

Largely as a result of the Oppenheimer trial of 1954, Robert Christy got involved in politics for the first time. Robert, one of the creators of the first atomic bomb (the "Christy Gadget"), became a strong public voice to oppose further atmospheric nuclear tests. President Eisenhower planned to continue and accelerate these tests. Adlai Stevenson, who was running again for the U.S. Presidency in the 1956 election, was opposed to atmospheric testing, both because of its health dangers and in hopes of reducing the Cold War tensions with the Soviet Union. Robert decided to publicly support Stevenson's bid for the U.S. Presidency.

The Los Angeles Times Gamble

Robert had already had arguments with Lee DuBridge, the President of Caltech, on the subject of atmospheric testing. Both DuBridge and the head of Physics, Bob Bacher, served on President Eisenhower's Science Advisory Committee. They were thus associated with Eisenhower's policies and supported the atmospheric testing. Robert became a spokesman for Caltech professors who were opposed to any further atmospheric testing of nuclear weapons. He and Tommy Lauritsen, a professor of experimental physics at Caltech and Robert's best friend until he died of cancer in the early 1970's, organized a group of a dozen professors to make a public statement in the Los Angeles Times. Robert recalled:

Adlai Stevenson had proposed that atmospheric testing of bombs should be stopped, because the radioactivity in the atmosphere was spread around the Earth and would (in a statistical way) cause a significant number of people world-wide to get cancer. This was not, you might say, a "danger" for any individual, because the hazards of everyday life were much more severe than the hazards presented by the atomic bomb. But statistically, considering a population of a billion people, there would be quite a significant number that would in the end get cancer as a result of these tests. This idea of the danger of tests had been promoted by Linus Pauling, among others. He travelled around and made public speeches pointing out the dangers to humanity as a result of this testing.

This [Pauling's warning of the dangers of atmospheric testing] was picked up by Adlai Stevenson, who made it one of his planks in running for President against Eisenhower at the time. On the other hand, many people supported Eisenhower, and Lee DuBridge was a prominent member of the President's advisory committee at the time, I believe. He was supporting Eisenhower, that the testing was not of any significant danger and it should be continued. So this, you might say, became a political debate. Pauling started it off as a humanitarian debate, and ultimately, of course, Pauling won the Nobel Peace Prize for his active participation in that debate.

A few of my colleagues at Caltech and I felt that the issue was important, and we felt that since we had some knowledge of the effects of nuclear radiations, that we should speak out publicly to support Stevenson's stand. Tommy Lauritsen and I were, I think, the prime movers in the subject. We recruited Carl Anderson, Joe Langmuir, and I think altogether about a dozen, but I do not remember all the names. We took out an ad in the L.A. Times (it may have been a full-page ad, I forget) to express our views.

The ad was funded by Robert Christy, Tommy Lauritsen, Carl Anderson, Joe Langmuir, and the other Caltech professors. Note that this committee was technically bipartisan as Carl Anderson was registered as a Republican at the time.

Robert conversing with a colleague (Caltech photo)

The Trustees of Caltech consisted of powerful people, most of whom supported atmospheric testing. They were in an uproar, and wanted Caltech President Lee DuBridge to control his faculty, including Robert Christy. Robert at the time was a tenured professor with immunity from being fired without just cause. Some of Trustees wanted to pressure

Robert with colleagues at Caltech (Caltech photo)

DuBridge to remove tenure from Robert and the other professors because they used the word "Caltech" in the L.A. Times ad. (It could be argued that use of the word "Caltech" falsely suggested that the ad represented Caltech's official policy, thus providing just cause to remove tenure.) Some Trustees even wished to fire the professors who had backed the ad, although most of them did not support so extreme a course. DuBridge just stayed quiet and did nothing to Robert and the other faculty — other than chewing them out for violating Caltech rules.

DuBridge risked his own neck, as the Trustees could have removed him for not following their guidelines. (In fact, in later decades one President and one Provost were removed from office for not following guidelines from the Trustees.) No matter what he may have thought privately, DuBridge was always one to defend people's rights to their own views — even though he felt it had been wrong to use the institutional name of Caltech in their private ad. Robert said:

> The fact that we mentioned that we were professors at Caltech upset the administration, who felt that Caltech's name should not be brought into such an issue. We were scolded by the president [of Caltech], and by the Board of Trustees — through the president: the Board of Trustees complained to the president, and President DuBridge scolded us. So it ended that way, that we were scolded by the president, but we still made our point.
>
> There was one Trustee who was exerting pressure to get Pauling fired. I don't think that they thought I was a big enough fish to worry about. But in Pauling's case, DuBridge stood up for the fact that Pauling had a right to express his views publicly, and DuBridge supported that right. The Trustee ultimately resigned.

It is noteworthy that a month later the Eisenhower government changed its stance on nuclear testing, and decided to eliminate atmospheric testing.

Adlai Stevenson lost the election, but the spirit of Robert's ambitious proposal continued, and eventually led to the Strategic Arms Limitation Treaty (SALT) talks.

Robert Discovers the Cause of Mysterious Communication Failures

During Cold War years in the 1950's, a number of mysterious communication disruptions occurred. It was feared that the communications had been sabotaged in some way by the Soviet Union. Robert was at Caltech at the time, but was also a consultant for the Rand Corporation, and became aware of this phenomenon.

For years Robert had been outspoken in his opposition to atmospheric testing of nuclear weapons, and had put a good deal of effort into understanding the effects. At that time the U.S. was still performing atmospheric tests of nuclear weapons. One test involved exploding an atomic bomb at a very high altitude, roughly 20 miles.

It had been known that atomic bombs could sometimes cause problems with electronics in the vicinity, but it was Robert who single-handedly worked out the physics by which atomic explosions in the upper atmosphere would produce an electromagnetic pulse (EMP) that could have catastrophic effects on circuits on the ground at very great distances, and could thereby disrupt communications. He was thus the first to connect the disruption of communications with the high-altitude nuclear explosions. He wrote this up as a classified report. It should be noted, however, that the warning in this report did not prevent the U.S. from carrying out the very-high-altitude "Starfish Prime" test of 1962. In this test a 1.4 megaton bomb was exploded over the Pacific Ocean at an altitude of 250 miles, causing electrical damage in Hawaii (about 900 miles away). The Soviets conducted similar high-altitude tests over Kazakhstan in the same year. These caused even more extensive damage since they were above an inhabited area rather than over the ocean.

The EMP effect of high-altitude atomic explosions is now widely known, but it was Robert Christy who first brought this phenomenon to the attention of the U.S. government.

Robert's Work Towards Initiating the SALT Talks

In the Cold War era the number of nuclear weapons worldwide escalated dangerously. After the mid-1950's, nuclear tests increased from a dozen or so a year to over a hundred in 1958 (over 70 by the U.S. alone), and nearly 140 in 1962 (nearly 100 by the U.S.). For about two decades thereafter roughly 50 nuclear tests were performed each year.

In the 1950's, Robert Christy and Tommy Lauritsen used every opportunity to propose arms limitation talks between the U.S. and the Soviet Union. However, the first Strategic Arms Limitations Treaty talks (SALT I) did not take place until 1969 through 1972. SALT II took place in 1977 through 1979.

Robert's Opposition to Higher-Tech Weapons in the Vietnam War

In 1966, Professor Jerrold Zacharias of the Massachusetts Institute of Technology (MIT) invited Robert to participate in a group to contribute to technological improvements in the weaponry used in the Vietnam War. Robert declined this invitation, stating that

> I actually am very doubtful if we can find a solution to our problem without some kind of revolution in our policy ... I do not believe a solution will be found by "technological innovations in military weapons and practices." Further, since all I see in that direction is more effective killing of people with whom I believe we have no basic quarrel, I am not even sympathetic to improving our techniques — I would almost certainly find that avenue one that I could not devote myself to.

Chapter 8

Becoming Oppenheimer's Successor at Caltech

Working with Fermi in Chicago, and Co-Habiting with Teller

After the Second World War came to an end, there was a big celebration at Los Alamos, and people started to look for peace-time jobs. For many of the scientists, this merely consisted of waiting until some university offered a position to them.

Robert recalled, "I was offered a job as an assistant professor at the University of Chicago. The University of Chicago was a first-rate institution, and Enrico Fermi was going there, so this was a very fine offer, and I accepted it and went to Chicago, I think, in approximately February of 1946." Robert taught a course there in the spring of 1946, and worked with Fermi on the design of the Argonne reactor, a precursor to the first power reactors (see Chapter 5). His salary was $5000 a year, roughly twice what he had been earning before at Illinois Tech and at the University of Chicago, and considerably more than he had been paid on the Manhattan Project at Los Alamos.

When Robert and his pregnant wife Dagmar arrived in Chicago with their toddler son Ted (age 1 ½), there were no single-family apartments to be found. No houses had been built during the war, so housing was almost impossible to find. Teller and his wife and young children encountered the same problem. All they could find available to rent was a mansion that was both too expensive and too big. Therefore the Christy and Teller family decided to rent this mansion together.

The Teller family used the mansion's kitchen as their kitchen, the living room as their living room, and some front bedrooms on the floor above for themselves. The Christy family used the butler's pantry for their kitchen, the dining room as their living room, and two back bedrooms above as their bedrooms. There was a separate servants' stair from near the butler's pantry to the back bedrooms, so although there was no actual wall separating the families, they did not have to go through each others' halls or rooms. It worked well.

When their second child Peter Robert Christy was born in May 1946, the story goes that he had to sleep in the bathtub.

Since housing was so difficult to find in Chicago, the Teller-Christy house became a temporary refuge for many of the scientists who arrived without anywhere to live. There were small servants' bedrooms on the third floor, and the visiting scientists would live there. Among these visitors was Stanley Frankel, an eminent chemist whom Robert had known from meetings at the National Academy, and his wife.

Robert recalled, "We had two separate housing arrangements in the same house but lived quite separately, basically. We lived there until some months later when I received a call from Willy Fowler at Caltech saying that they wanted me to come to Caltech as a professor."

Leaving for Caltech

After the war, Oppenheimer had gone back to his joint positions at Berkeley and Caltech. He had been the theoretical advisor at the Kellogg Radiation Laboratory at Caltech, a group of experimental physicists headed by Charlie Lauritsen.

However, soon after arriving at Caltech, Oppenheimer left for Washington, DC to become a national spokesman for science, to try to initiate policies to control nuclear materials for peaceful purposes and stifle a nuclear arms race. He was on the Board of Consultants to the committee (appointed by President Truman) that produced the Acheson-Lilienthal Report, advocating creation of an international Atomic Development Authority to own all fissionable material, all uranium mines and related laboratories, and all nuclear power plants. This resulted in the Baruch Plan, a proposal to the United Nations; however, it was rejected by the Soviet Union, which objected to the provisions for the inspection of all the Soviet Union's uranium resources (such as the mines in Czechoslovakia, which had previously been controlled by Hitler). It had become clear that a nuclear arms race was starting.

As Robert recalled, "Robert Oppenheimer, who had taken a position [at Caltech], had found that he was spending so much of his time in Washington and going back and forth that it was just not feasible, so that he wanted to move." Charlie Lauritsen, the head of Caltech's Kellogg Radiation Laboratory, thus had to look for a new theoretical advisor to replace Oppenheimer.

Lauritsen asked Oppenheimer for a recommendation, and he promptly suggested his former student Robert Christy who at that time was on the University of Chicago faculty. Oppenheimer called him "one of the best in the world." Charlie Lauritsen decided to invite Robert Christy to be Oppenheimer's successor at Caltech, though as a full-time position rather than the part-time one that Oppenheimer had had. William A. Fowler — Willy to all who knew him — was chosen to call Robert and try to persuade him to come to Caltech. Robert told Willy that he would consult his wife. (Robert himself would have been O.K. with either staying at Chicago or moving to Caltech — both had excellent physics departments.) Robert went home and asked his wife Dagmar which position he should choose, and she answered, "Caltech." So they went to Caltech.

Early Life at Caltech

Robert visited Caltech on his own to find housing prior to the move, taking a plane from Chicago to Los Angeles. It was the first plane trip of his life, on a propeller plane that stopped at cities en route every few hundred miles. It flew at 10,000 feet, just above the turbulent atmospheric layers; but every time it descended to land the turbulence caused Robert to feel sick. He found it a very unpleasant trip.

There were no apartments to rent in Pasadena, so he ended up having to buy a house about a mile from Caltech, at 175 S. Greenwood (between Del Mar and Colorado). As Robert put it, "This cost $12,000, a fantastic sum of money, considering that I had nothing." (His future annual salary at Caltech would be $5400 a year, up $400 from the Chicago salary.) To enable him to buy the house, Caltech loaned him the money for the down payment, which he would not otherwise have been able to afford.

Robert and his family travelled by train to Caltech in September 1946. This was, as Robert put it, "very, very comfortable" compared to the

Robert Christy in the garden of his house in Pasadena

plane trip. Robert remembers that their son Peter, who was only a few months old, was kept safe in a basket on an upper berth.

Life was also challenging because they could not at first afford a car. All travel was done by bicycle, which also was used to carry the groceries. For greater distances buses or taxis had to do. After one year Robert was finally able to buy a ten-year-old car, a 1937 Chevy.

Large Numbers of Graduate Students

There was nobody appropriate at Caltech to teach theoretical physics courses to the graduate students: all the other physics faculty, including Charlie Lauritsen, his son Tommy Lauritsen, and Willy Fowler, were in experimental physics. Therefore Robert ended up teaching *all* of the graduate physics theory courses.

Robert was also beset with graduate students who wanted to do their theses. A thesis project under the supervision of a professor is how a graduate student learns to perform original research (rather than just solve the problems at the back of a textbook). Because the student lacks the professor's experience and generally has not had time to build up an extensive knowledge of the scientific literature, supervising a graduate student can be fairly time-consuming — though this depends to some extent on the quality and experience of the student and on how difficult the project turns out to be. A recent check of Ph.D. theses of Robert's students (stored in his office at Caltech) yielded the following numbers: three completed in 1948 and 1949, nine in the 1950's, five in the 1960's, and one in 1970. This can only be an underestimate, as some of his favorite theses were kept at his office at home and others may be missing.

Between teaching and supervising, Robert had very little time to carry out research on his own. As he recalled,

> One of my main jobs was of course teaching classes. Basically, there was no one else there to teach modern theory at all, and so I had the responsibility for a sizable portion of the graduate teaching: I was quite busy teaching. Also, there were a number of students who had accumulated, and had been unable to finish up theses before and during the war, so there were people there who wanted to do theses and get degrees. With one thing and another, I was fairly busy.

Robert was highly valued by Caltech, as is shown by the letter below. He was considered one of the "key professors" of Caltech, to whom they granted an unusually large salary increase.

CALIFORNIA INSTITUTE OF TECHNOLOGY
PASADENA

OFFICE OF THE PRESIDENT

July 9, 1954

Professor Robert F. Christy
2810 Estado Street
Pasadena 8, California

Dear Professor Christy:

       Recognizing the great importance to the Institute of
maintaining an adequate salary scale for its key faculty members,
Mr. Keith Spalding has recently made a very generous arrangement
to provide an increased income to the Institute which will allow
giving substantial salary increases to a very small number of key
members of the faculty. The faculty members judged to have rendered distinguished service to the Institute such as to merit special
recognition were chosen upon recommendation of the Chairmen of the
Divisions and the Dean of the Faculty. They were given final ap-
proval by a special committee of the Board of Trustees.

       I take pleasure in informing you that you have been
selected as one to receive such an increase. Consequently, your
salary, beginning July 1, 1954, will be raised to $10,000 per year.
You will realize that the Board of Trustees and Administration are
pleased indeed that, even in a year of some financial stringency,
the generosity of Mr. Spalding has made possible a group of critical
salary increases which are not only richly deserved, but long overdue.

Sincerely yours,

L. A. DuBridge
President

LAD:fh

cc: Dean of the Faculty
    Central Files
    R. F. Bacher

Robert was always interested in experiments, and would hang around the lab
at times, but his viewpoint was that of a theorist. He was unusual as a theorist in
that his greatest strength was not in creating new theories (which is what theorists
like Feynman and Schwinger were very good at), but rather in seeing how theory
and experiment related.

J. Robert Oppenheimer is at front left, with Lee DuBridge just behind him at left; Robert Millikan is fourth in the front row, Charlie Lauritsen is fourth from right in next-to-last row, and Robert Christy is at center of back row. (Caltech photo)

Robert's Cosmic Ray Work

Robert spent some time continuing to study cosmic rays, as he had done at Berkeley. In his words:

> My work initially was on things related to cosmic rays and elementary particles. This was what I had done before the war for my thesis, and I continued to have an interest in these fields. That was where the main work on elementary particles was being carried out, in connection with cosmic rays, because there were no high-energy accelerators to make new particles at the time — those were just being developed.
>
> I remember one important issue that I got involved in. That was the so-called "two-meson hypothesis." I mentioned that my thesis research was on the burst production by the mu meson (as it is now known — we called it the mesotron at the time). There were contradictions. There was a meson that was theorized to be involved in nuclear forces. These were copiously produced in cosmic rays, and if they were produced they had to be interacting with something; and yet when you saw them going through the atmosphere, they appeared not to interact [via the

strong nuclear force] at all. They penetrated deep down into the Earth without interacting except by the electromagnetic interactions that I mentioned in my thesis. Thus the behavior of the mesotron was a puzzle. The puzzle was that it was made readily, and yet did not interact [strongly]. This puzzle was clearly present in the late forties and many people were trying to understand it. The clue came from experiments done with photographic emulsions flown at very high altitude by, I think, a British team — I remember one of the names was Powell. They found evidence of nuclear reactions (which showed up in these photographic plates) making particles, and then a particle stopping and decaying — changing its nature into another particle — and then that other particle penetrating. The idea began to appear that what was going on was that there were two mesons: that the mesotron that we saw at sea level was not strongly interacting, but that it was made by the decay of a strongly interacting particle. This was the two-meson hypothesis.

In order to prove that, one had to be able to show in detail the way that worked out. I had an idea for that and I assigned a student, Richard Latter, to work on that for his thesis [because I was too busy to have time to work on it myself]. But students take a long time to do things so it took him a couple of years to work this out. In the meantime, Robert Marshak at Rochester and a student working with him had the idea, and they worked it out and published it long before my student finished his thesis. I always regretted the fact that I was not able to share in that discovery, simply because I had assigned it to a student to do as a thesis.

Richard Latter finished this project and published it in his thesis in 1949. Robert Christy sadly recalled, "As I say, this was one of those occasions where having extra students to look after slows you down a bit."

There were other projects, a few of which were circulated as preprints among his colleagues but never actually published in a journal. Robert says, "I remember one thing that I wrote on the stability of orbits in the synchrotron that didn't get published, but I don't remember other things."

The Growing Mistakes Paper

In those days, papers were written by hand: they were truly manuscripts. After Robert had finished writing a paper, he would give it to a Kellogg secretary to type.

There was one particular paper where this process did not work. The secretary typed the paper and returned it, and as would sometimes happen, Robert found a few mistakes in it. He corrected the mistakes and gave it back to her to re-type. She did so, but this time the re-typed paper had more mistakes than before. This back-and-forth continued several times, with more mistakes creeping in each time the paper was re-typed.

Finally Robert gave up. This paper was never submitted for publication.

Discovery of Which Nuclear Reactions Actually Power the Sun

In the early 1950's, a priest, Father James O'Reilly, came to Robert as a graduate student wanting to get a Ph.D. in astrophysics. Robert gave him the problem of working on the Sun's nuclear energy source. In Robert's words:

> I had been with a number of the other people at Kellogg and taken some seminars in stellar structure, and I thought it would be interesting to evaluate the energy production in the Sun and find out which of the possible competing reactions in fact was the dominant one in the Sun. The competing reactions were the proton-proton chain and the carbon cycle [CNO cycle]. The carbon cycle involves a cycle in which protons and carbon, by a series of reactions, make helium [with the carbon acting essentially as a catalyst — the carbon remains, after the four protons have been converted into a helium nucleus]. The proton-proton reaction is a direct combination of two protons to make deuterium, and then further additions of protons makes tritium and helium-3, finally producing helium-4. The reaction constants were fairly well known, so I had Father O'Reilly calculate the interior structure of the Sun in order to find out which reaction was most important [under solar conditions].

Hans Bethe, who had been Robert's boss at Los Alamos, had already worked on that problem in the late 1930's with Carl Friedrich von Weizsäcker. They had explained the Sun's energy as being due to the CNO cycle. Robert decided to revisit the subject, and together with Father O'Reilly, discovered that in fact the Bethe and von Weizsäcker had been wrong. The CNO cycle was *not* responsible for the Sun's energy generation. It was the proton-proton chain that powered the Sun, with only a tiny contribution from the CNO cycle.

The CNO-cycle reactions are more temperature-sensitive than the proton-proton chain. Therefore the CNO-cycle dominates in stars more massive than the Sun, which have higher temperatures in their central cores. The temperature in the Sun's core is slightly lower, and the proton-proton chain is dominant there.

Lee DuBridge and Bob Bacher

In the fall of 1946, very shortly after Robert had arrived, Lee DuBridge came to Caltech as its first "real" President. (DuBridge's predecessor, Robert Millikan, did not call himself president, but only the "Executive Head.") In 1949, after some arm-twisting by Lee DuBridge, Bob Bacher came from Cornell to Caltech to serve as the Chairman of the Division of Physics, Mathematics, and Astronomy.

Bob Bacher had been at Los Alamos, where he had been the head of the Experimental Physics and Bomb Division, and he and Robert Christy had known each other there. Robert had not known DuBridge, but Bacher and DuBridge had already been friends and colleagues. In 1940 Bob Bacher had joined the Radiation Lab at MIT, where he headed the radar receiver group under the overall direction of Lee DuBridge. The two of them got along very well with each other.

Luring Feynman to Caltech

One of the first acts of Bob Bacher as Division Chairman was to lure Richard Feynman to Caltech. After Los Alamos, Feynman had declined an offer from the Institute for Advanced Study at Princeton, New Jersey, despite the presence there of such distinguished faculty members as Albert Einstein, Kurt Gödel, and John von Neumann. Instead Hans Bethe had taken Feynman (who had been one of his group at Los Alamos) with him to Cornell. Feynman was very happy there. One of the problems he had worked on there was analyzing the physics of a twirling, nutating disc as it is moving through the air — this was inspired by an actual thrown plate that he had observed in the cafeteria at Cornell. This work, which used equations of rotation to express various spinning speeds, proved important to the work he did at Cornell in the late 1940's on quantum electrodynamics, for which he was awarded the Nobel Prize in 1965.

While still at Cornell, Bacher had interacted further with Feynman. He had heard that Feynman might be considering taking a sabbatical to which he was entitled, in order to go to Brazil for a year. As soon as Bacher was at Caltech as Division Chairman, he began the attempt to lure Feynman to Caltech. This was a complicated maneuver. First, Bacher had Feynman visit and stay at his home right next to the Caltech Campus. Secondly, Bacher loaned Feynman his own car to go anywhere he wanted, and what Feynman wanted was to go to see the entertainment of Hollywood. In addition, Bacher offered Feynman a sabbatical year as his first year at Caltech, since Feynman wanted to go on sabbatical to Brazil — normally, a sabbatical year is granted only after seven years of work. Finally, at Cornell Bethe had spoiled Feynman by taking on all the graduate students who otherwise would have landed with Feynman, who did not want to be obliged to take on graduate students. Bacher came to Robert and said, "Well, what are we going to do?" And Robert said, "Well, I guess I could agree to supervise the theoretical students' thesis work." Robert recalled these events:

> I don't remember the exact date when he [Dick Feynman] came. I do remember that Bob Bacher proposed that we bring him here. I and everyone else around were certainly very much in favor, because it was clear that he was a remarkable and in fact unique physicist. Bob Bacher kept working on Feynman, writing to him and trying to attract him here. Finally Dick agreed to come, but there were a number of conditions. One was that he had a sabbatical coming up, and he had planned to go to Brazil on his sabbatical. The condition that Bacher met was that in the first year of Feynman's appointment he would not in fact be here; he would be on sabbatical. He got his appointment here and immediately took off on sabbatical to Brazil because that was promised him, that he would not lose that year in Brazil.
>
> Another condition, I remember, was that Dick explained that he didn't want to be responsible for graduate students' theses, because he wanted to be able to proceed to do research at his own pace, and graduate students would be a drag on that. This doesn't mean that he didn't take any graduate students, but he would take them only when he chose, when he wished to. He did not want to feel obliged to take them. At that point I agreed that, in order to get Feynman here, I would be responsible for all the graduate students. That is, Dick would work with them if he wanted, but if he didn't want to, I would be responsible for supervising their theses. That made extra work for me, but it was worth it in order to have Feynman come here.

This offer by Robert to take on Feynman's graduate students in addition to his own is a striking example of Robert's generosity. It was a major sacrifice on Robert's part. It would eat up almost all of the remaining time that he could have spent on doing the theoretical physics research that he enjoyed so much and that he had proven to be so good at under Oppenheimer, Wigner, and Fermi.

At least nine graduate students in the 1950's obtained their Ph.D.'s under Robert's supervision. Robert says, "Most of my research ideas I fed to graduate students, because I did not have time to proceed on my own independent research. I was so busy trying to keep graduate students busy." Robert took on this task because he wanted Caltech to have the very best people possible.

Robert Christy and Richard Feynman had a great respect for each other, and had become friends while at Los Alamos. Between the inducements and this friendship, Feynman decided in the end to come to Caltech.

From left: Richard Feynman, Robert Christy, Bob Bacher, and Sin-Itiro Tomonaga — who later shared the Nobel Prize with Feynman (Caltech photo)

The reason why Feynman did not like to supervise graduate students was that he felt that by the time he had posed the question sufficiently clearly to offer to a graduate student, it was at the stage where he could solve the problem in an evening himself — and if he could get a problem that clearly set, he could not refrain from doing it himself. On the other hand, Feynman *did* enjoy teaching. As Robert recalled:

> He was an excellent teacher, in fact he wanted to teach. He taught a great variety of courses, and he taught them in his own unique way.
>
> One of the courses he decided to teach was the freshman course in Physics, or the freshman-sophomore sequence in Physics. The result of that was the famous Feynman Physics volumes. It turned out that they were an excellent course in physics, but reviewed by most people as being too difficult for the sophomores and freshmen, but excellent for the faculty.
>
> He also taught many other courses, which he would again do his own unique way. He started a course called "Mathematical Methods in Physics," and after he had taught that course for a while then other people taught it; but he initiated it. He was an excellent teacher, but he essentially wanted to do things according to his interests rather than feel he had to do them.

Over a decade later, in January of 1962, Robert was invited to nominate someone for the Nobel Prize in Physics. He nominated Feynman. In his nomination letter, Robert wrote:

> In response to your invitation to nominate a candidate for the Nobel prize in physics, I offer the following:
>
> Richard P. Feynman
>
> for the invention and discovery of Feynman Diagrams for performing calculations in quantum field theory and particularly for calculations in quantum electrodynamics.
>
> I believe that this invention of Feynman's has had a major effect on most of the subsequent work in this field. This method has so facilitated calculation, the many calculations now made would be too unwieldy without the diagram method. The universal adoption of the method is sufficient demonstration of its importance.
>
> I will not here document his published works in this field since you no doubt have these references already. I wish primarily to emphasize the importance of this contribution to the further development of theoretical physics over a period of many years.

Feynman was awarded the Nobel Prize... but not until 1965.

The Vista Project at Caltech: the Defense of Western Europe

In 1951, a secret project was organized by Caltech to consider how to defend Western Europe against the Soviet Union. Lee DuBridge, a physicist and the first President of Caltech, convinced the Trustees and administration to support this "Vista Project." It would bring a lucrative government contract to Caltech, and perform a valuable national service at a time of great international tension.

About a hundred scientists participated in the project, headed by William A. Fowler. Robert Christy was a participant; among the others were Lee DuBridge, Robert Oppenheimer, Charlie and Tommy Lauritsen, and Jesse Greenstein. It took place in the La Vista del Arroyo Hotel in Pasadena, on the east bank of the arroyo. The project lasted for about a year. It was a full-time job for Robert Christy, who was on leave from Caltech — although he still had to supervise his graduate students. Robert recalled this project:

> Project Vista was organized sometime around 1950 or '51 at Caltech, primarily by Charlie Lauritsen (and others; but the Caltech organization was primarily by Charlie Lauritsen and Lee DuBridge). The project was intended to provide advice to the government on the defense of Western Europe against a possible invasion

by the Soviet Union. This was a matter of great concern at the time, and the state of the defenses in Western Europe was such that the Soviet Union could have over-run Western Europe without too much difficulty. One of the ideas in this was to see whether nuclear weapons could be used in the defense — tactically, on the battlefield — to prevent such an onslaught. There were many other technical questions that were examined too, in the course of this.

People were recruited from quite a number of institutions. We occupied an old hotel, the Arroyo Vista hotel on the banks of the arroyo, and so it was called Project Vista. This went on for approximately one year. Willy Fowler was made the director. He was an excellent director, he ran things very well. Charlie Lauritsen was, you might say, the *eminence gris* behind the scenes, advising the director.

This led me to spend a year. I travelled around with others to many army and air force bases to learn how military operations worked, how they were conducted, and to learn things about the armed services that I had not known. It was a period of a lot of education as far as I was concerned, and I did make some contributions on the use of tactical nuclear weapons on the battlefield.

Some of the members of the Vista Project: Robert Christy is second from left, and Chairman Willy Fowler fourth (Caltech photo)

The result of this was a report which went to the government, and was largely suppressed. The reason it was suppressed was that the nuclear weapons up to that time had been the sole prerogative of the Strategic Air Force (which did long range bombing from behind the battle lines). We were proposing the use of atomic weapons on the battle front. At that time there was a shortage of nuclear weapons — there were not very many. The Strategic Air Force felt that if the supply of U-235 and plutonium was diverted to the battlefield weapons then they would not have so many for the long-range bombing, so there would be a competition between different services. The Strategic Air Force, at the time, was the one that controlled the nuclear weapons, and they did not want competition from the battlefield for nuclear weapons. So they suppressed the report, which could probably barely be found in government archives at this point; but I'm sure Caltech has a copy in its archives — I would assume so, at least, although it was a secret report, so I don't know where it sits.

That took about a year and I learned a lot, but it also takes a year out of your life at a time when physics is very active, so it doesn't promote your physics very much to spend a year doing that. There is a down-side to all these things.

More members of the Vista Project. Chairman Willy Fowler is the third person in the back row, and Robert Christy is the fifth (Caltech photo)

Robert's Work as a Consultant

Robert had contracts to do consulting for a number of companies over the years. We found documents of such agreements in the 1950's with the Atomic Energy Commission for consulting at the University of California, with the Rand Corporation, with the Union Carbide Nuclear Company, and with the Los Alamos National Laboratory.

He continued to do consulting work for several decades thereafter, until he was 86 years old. His largest consulting project was for the joint US-Japan dosimetry reassessment described in Chapter 16.

Chapter 9

Contributions to Physics at Caltech

Most physicists are either experimentalists or theorists but not both. Robert was the one of the few exceptions. He always excelled at both.

The Kellogg Radiation Laboratory at Caltech was largely devoted to experimental physics. Robert was known to spend considerable amounts of time in the labs, helping to resolve problems with the equipment. The experimental scientists working there considered him one of their own. However, he considered himself really to be a theoretical physicist, and came up with a number of important theoretical contributions.

Since it was the experimentalists of Kellogg Radiation Laboratory who had hired Robert to come to Caltech, he felt that he owed them something. This

Willy Fowler, Richard Feynman, Niels Bohr, Bill Houston, Pellam, Bob Bacher, Robert Christy, Lee DuBridge, Carl Anderson, and Tommy Lauritsen at Caltech (Caltech photo)

149

contributed towards his willingness to work on some of the problems they had with their experimental apparatus. After about a decade, to get more time for theory, he decided to finally move out on his own by taking an office in Sloan Laboratory. There he would not be quite so easily accessible to the experiments — and to the experimentalists.

Tommy Lauritsen, Max Delbrück, Niels Bohr, and an unidentified person at Caltech (Caltech photo)

Charlie Barnes was one of the professors in experimental physics at the Kellogg Radiation Laboratory. In a recent letter he wrote:

> Of course, you know how much I admire and respect Bob [Robert Christy]. I have known him since 1953, when I first came to Caltech (from UBC in Vancouver, B.C.), and he was such a great help to me, and indeed to all of us in those old days. ... Another recollection of Bob's great versatility was that he built a large swimming pool at a house he used to live in, largely with his own hands. This was not an amateur's effort; it was a very professional looking effort, and Bob tied all the steel together before the cement trucks came and piped the concrete in by big hoses. I don't know any other theorist who could do that and get it done right. In fact, only a few experimentalists could have done it. Well, this is just one more of Bob's many talents!

In another letter Charlie Barnes talked in more specific terms of the contributions that Robert had made to experimental nuclear physics at Kellogg:

Robert at one of the many Kellogg parties (Caltech photo)

While in the Kellogg Lab, Christy came into the accelerator labs essentially daily, to see what new information was emerging about the nuclear physics of the light nuclei (atomic number less than 20), which the lab had adopted as most interesting and best suited to the Lab's low-energy accelerators. With his knowledge of experimental physics, he frequently asked questions about the techniques being applied, and sometimes made clever suggestions about how these might be improved. At the same time, with his daily checking of the results, he played an important role in suggesting alternative or additional experiments that improved the rapid progress being made.

Bob Bacher, Niels Bohr, and Lee DuBridge (Caltech photo)

Eventually both the Kellogg Lab and the Division of Physics, Mathematics, and Astronomy grew considerably. More professors were hired, and Robert was no longer a lone theoretical physicist among experimentalists.

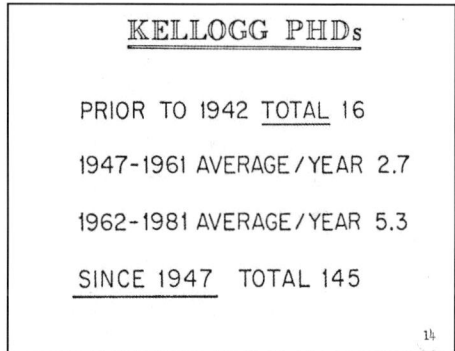

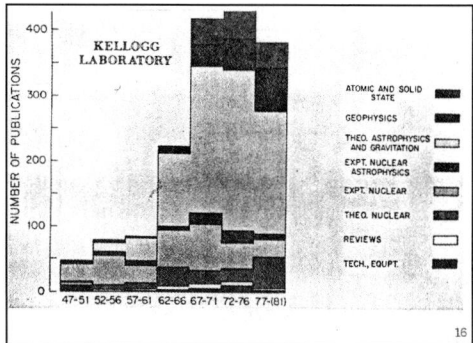

KELLOGG PHDs

PRIOR TO 1942 TOTAL 16

1947-1961 AVERAGE/YEAR 2.7

1962-1981 AVERAGE/YEAR 5.3

SINCE 1947 TOTAL 145

14

KELLOGG RESEARCH FELLOWS
VISITING ASSOCIATES ......

BEFORE 1950 TOTAL: 8

1950-1959 AVERAGE 4/YEAR

1960-1981 AVERAGE 14.7/YEAR

SINCE 1950 399 SUCH APPOINTMENTS
HAVE BEEN MADE BY KELLOGG

15

KELLOGG LABORATORY

NUMBER OF PUBLICATIONS

400

300

200

100

0

47-51 52-56 57-61 62-66 67-71 72-76 77-(81)

ATOMIC AND SOLID STATE
GEOPHYSICS
THEO. ASTROPHYSICS AND GRAVITATION
EXPT. NUCLEAR ASTROPHYSICS
EXPT. NUCLEAR
THEO. NUCLEAR
REVIEWS
TECH., EQUPT.

16

Three slides from a Kellogg talk (from Robert's files — though he may not be the author) illustrate that the Kellogg Radiation Laboratory became a much larger group in the 1960's

Attending a physics colloquium. Front row: Willy Fowler, Bill Houston, Lee DuBridge, Bob Bacher, Niels Bohr, Robert Christy, Richard Feynman, and Max Delbrück. (Caltech photo)

The Necktie Story of R. F. Christy

The following anecdote, of an incident from this time period, was provided by Professor G. J. ("Jerry") Wasserburg of Caltech:

> Bob Christy always wore a Navajo-made thunderbird tie. He wore it with a tie clasp. As time passed, the tie began to look rather worn, particularly where the tie clasp held it. Bob wore that same tie with great regularity. At the Athenaeum [a club at Caltech where faculty frequently met for meals], it was surmised that he did not own another one.
>
> He was sometimes lightly teased about it. During the late 50's I was working in New Mexico and Arizona, occasionally with visits to Los Alamos: very much

in the Albuquerque area. I purchased a new Navajo-made thunderbird tie outside of town with the idea of giving it to Christy as a teasing joke. Around some Christmas season, I do not remember the exact year, I arranged with his secretary to have it placed in his mail box with a note "from an admirer." Well, in a short time Christy appeared at the Athenaeum wearing the new tie. We frequently had lunch together and walked back to our respective offices. I always thought we looked like Mutt and Jeff when walking and talking together [since Christy was much taller].

After he began appearing with the new tie, I remarked, "Bob, I see you got a nice new tie. What brought that on?" He replied, "An admirer gave it to me." No further development of the issue came from Bob. After a while, I was walking with him again and asked, "Well, who is the admirer that gave you the tie?" "I don't know," was the answer, and no further delving seemed to pique his curiosity. Finally, one day I said to him, "Bob, I am the person who gave you the tie, with the hope of stirring you up." His answer, true to form, was, "Thank you Jerry, someday I may do something for you." That got my dander up and I replied, "Well, Bob, you can give me your old tie." He answered, "No, but I will sell it to you for a dollar." So I gave him the dollar, and the next time we met, I got the tie.

Now what in the heck was I going to do with it! So, I took a thumb tack and mounted it on my bulletin board where I exhibited interesting oddments. After some years, more space was needed on the board, and I put the tie in a filing cabinet in

The framed necktie

which I stored mass spectrometer charts. It was then buried under the charts. After a few years, Dimitri Papanastassiou (who had been working in Kellogg) came to do his doctorate work with me. I needed some data on Sr [the element strontium] that had been done by a student who had long since graduated, and asked Papanastassiou to look through the file drawers. After some time I heard "Professor Christy" uttered. Papanastassiou had come across the tie, and instantly recognized who the wearer had been, saying, "He wore that tie in all the lectures that I took from him. That is Bob Christy!"

The personification of the tie made me understand what a true treasure I had. But what to do with this valuable memento? I decided to get it framed with what appeared to be a white shirt in a long, thin frame. The picture framer thought I was mad, but so it is. The long thin frame of the Robert F. Christy tie and shirt was then hung on the tall wooden door frame at the appropriate height (looking down!) and stayed there as a symbol of Christy until I moved out. It now is in the Caltech archives.

—Jerry (G. J. Wasserburg).

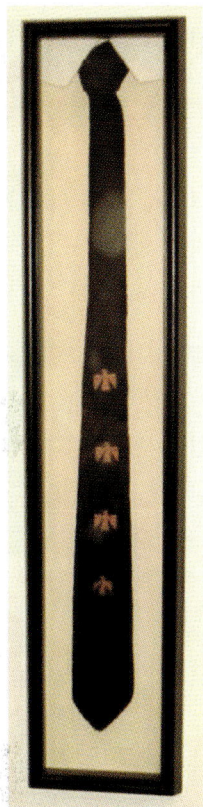

(Caltech photo)

Front

BACHER	ANDERSON	
NEHER	BETHE	
FOWLER	SMYTHE	
DAVIS	GOETZ	
LAURITSEN, C.	GELL-MANN	
DuMOND		

Middle

WHALING	SYMONS	
NE'EMAN	KING	
VOGT	SUTTON	
BURBIDGE	DEERY	
SEEGER	MATHEWS	
TOMBRELLO	MULLINS	
BARNES		

Back

van PUTTEN	COWAN	
KOBRAK	BOURY	
STRONG	IBEN	
HILDEBRANDT	EICHLER	
SWANSON	HARA	
WAHLBORN	ALEXANDER	
CHRISTY	HUNT	

Above: Front row: Margaret Burbage, Jeff Burbage, Fred Hoyle; at right (end of second row) is Robert Christy, dozing off wearing the thunderbird tie. Below: dozing again. (Caltech photos)

The Orion Spacecraft Project

Not quite all of Robert's research took place at Caltech. One summer Freeman Dyson lured Robert to La Jolla to work on the Orion spacecraft project for several weeks.

The Orion Project was initiated in 1957 and continued for several years. It was led by Freeman Dyson, an eminent physicist from the Institute for Advanced Study at Princeton, and Ted Taylor from General Atomics. At Taylor's request Dyson had taken a year's absence to work on this project. It was a study of the possibility of using atomic explosions to propel a spaceship, which could be very large. Repeated atomic explosions behind a "pusher plate" at the back of the spaceship would provide propulsion. As Robert put it:

> Freeman Dyson had an idea at the time (I forget the period) to develop a huge spaceship that could be sent off to Mars and places [the Orion Project]. He used to talk about this huge spaceship as weighing about as much as a destroyer would weigh, about five thousand tons. It would be propelled by setting off hydrogen bombs that would be sent out from the rear. They would explode, and the explosion would push the ship until it attained very high velocities.
>
> This sounds like a mad project, and it probably was, but I worked on it with Freeman for a period — I forget whether it was two weeks or six weeks — down in La Jolla. He was centered, I think, at General Atomics at the time. It was a nice place to go with the family for a summer and spend some time down there and assist in his project. Nothing ever finally came of this, although it's possible that the project was proved feasible — I'm not sure about that. But it was a very preliminary and wild idea.
>
> I always used to enjoy Dyson. He was an imaginative person with a very wide and thorough knowledge of physics. So it was an enjoyable thing to work on for a few weeks.

The Orion spacecraft design appeared to be technically feasible as a method of accelerating a large spacecraft to much higher velocities than could be achieved with chemical rockets. A prototype was built using conventional explosives, demonstrating that the concept was indeed practical. However, there are major disadvantages associated with having to explode nuclear bombs to propel the spaceship. Besides the question of maintaining security for these bombs, one would not wish to use this method of propulsion near the Earth where the electromagnetic pulse (EMP) from the bombs could affect electronics. This latter problem was one that Robert was well aware of since he was the first one to discover it, in his classified work in the 1950's.

In the end the Limited Test Ban Treaty of 1963 (also known as the Partial Test Ban Treaty) put an end to the Orion Project. This treaty prohibited nuclear tests in the atmosphere, outer space, and underwater. Both Robert Christy and Freeman Dyson had supported such a nuclear test ban. Of course this treaty also meant that the Orion spacecraft would be prohibited since it would have required nuclear explosions in space.

Robert honors Willy Fowler in a memorial talk, in 1995. (Caltech photo by Bob Paz, Institute Photographer, in *Engineering & Science* 1996, No. 2, p. 34.)

Chapter 10

Pioneering Work in Astrophysics: Variable Stars

Robert's First Sabbatical — with Oppenheimer at Princeton

In 1960 Oppenheimer, the Director of the Institute for Advanced Study at Princeton, invited Robert to come there for a year. Robert was very pleased. He accepted the invitation, although he only stayed for nine months. His teenage sons were upset about leaving their school and their friends. Nevertheless the family headed off to Princeton in Robert's car.

At Princeton, Robert had originally proposed to look into certain aspects of particle physics (multiple meson production in high-energy collisions, and structure of elementary particles). However, he started going to the library and reading a great deal in astronomy, mostly papers in the Astrophysical Journal, and some textbooks. This was a special pleasure for him, because at Caltech Fritz Zwicky, an eminent but rather eccentric astronomer, had banned physicists from entering the Astronomy Library.

Robert was interested in learning about the flow of radiation through the atmospheres of stars. Computers were just beginning to be available then for computations, and Robert began to write programs to teach himself programming — he was still impressed with what Feynman had done with the calculating machines at Los Alamos during the Manhattan Project.

Robert Christy (Caltech photo)

Back at Caltech: Diving into Astrophysics

After Robert returned to Caltech in 1961 he continued to immerse himself in this field of science that was totally new to him — theoretical astrophysics. As he had done before when his theoretical work on muons was guided by cosmic ray observations, his new work in astrophysics was guided by observations of variable stars, in particular the Cepheid and RR Lyrae variables. These stars are crucial because they can be used to provide distances to objects far out in space, not only in our own galaxy but also in other galaxies. There is a relationship between the length of the period of the variability of these stars and their intrinsic luminosity.

Most changes in stars take place on immensely long timescales, impossible to observe in a human lifetime. In contrast, the RR Lyrae stars vary with a period of hours, and the Cepheids vary with a period ranging from days to weeks. In addition, these stars are intrinsically very bright stars. They are thus relatively easy to observe. A wealth of observations of such stars was available, stretching over many decades. However, the reason for their variability was completely unknown at that time.

Robert's hope was to clean up this subject: not only to understand the basic mechanism that was driving the variability of these stars, but to provide a theoretical understanding that would allow him to make detailed predictions of the key properties of these stars. Up until then, theoretical calculations had only been able to handle very small variations (via first-order linear perturbation theory). This was not applicable to these stars, which exhibited major variations. Robert set up equations and computer programs that could handle such large variations, building computer models of the hydrodynamics and radiation transport in the outer envelopes of stars. This was based on the full set of non-linear equations that described the physics of these regions. This bore some similarity to the work that he had done on the theory of large explosions while on the Manhattan Project. In particular, consideration of shock waves turned out also to be crucial for the theory of these variable stars.

This was the time when the first computers became available for use at the university level, allowing much more complex calculations to be performed once one had learned how to program them. So Robert learned computer programming by himself. At the time he was doing this work he still had his large load of teaching, the double dose of graduate students to supervise (his own and those who would have been Feynman's), and all the committee work of a university professor. In Robert's own words from his 2006 interview with Sara Lippincott:

> It was unknown why they [Cepheids] varied, what made them vary. It was known that they were apparently spherical pulsators. That is, they expanded and contracted — a regular expansion and contraction — in spherical symmetry. That was known observationally, but why was it so?
>
> I knew something about the static structure of a star because I'd studied that a little bit. And I thought: Well, this is very much like the spherical hydrodynamics in implosions. It's basically the same equations we had used — of course with different substances — but the mathematical approach was very similar to what we had been working on at Los Alamos. I had never done those implosion calculations myself at Los Alamos — as I told you, I was a kind of go-between — but I knew about them. So I started to apply that theory to stars, and to apply the numerical techniques that had been developed at Los Alamos to this problem on RR Lyrae stars, which are similar to Cepheid variables but smaller. The approach I used was, I set up a mathematical model of the star and calculated its motion. I started the star moving by imposing initial conditions that had it collapsing a little bit — just a little bit away from equilibrium. And then, of course, it had to rebound. I started it vibrating and the model kept on vibrating back and forth.
>
> The key question was, could I make a sufficiently efficient program so that I could do this in a reasonable time on the computers then available, and what would happen when I started the star vibrating? Would the amplitude of the vibration slowly diminish and stop — which is how any normal thing would behave: you start it vibrating, and then the vibrations die away and stop. But in the case of these stars, I found that if I picked the right stars and started them vibrating, the amplitude of the vibrations would increase. And they increased until they reached approximately the amplitude that is seen in the sky for these stars. And there the amplitude leveled off and the model kept on vibrating forever: it was a self-excited oscillation. That is, even if I hadn't started the model vibrating it would have found some little quiver and gradually would have built up an oscillation. I started it to save time, but then the amplitude increased by itself. Of course the source of energy was simple enough. A star is a fantastic source of energy; the energy comes from the center of the star and flows out. So it has all sorts of energy available. The question is, how does it use this energy to make vibrations?
>
> So I studied these calculations and was able to understand why such stars vibrated. I was able to show that only stars in a particular region of the color-magnitude diagram would vibrate in the way that was seen. And I was able to get relations, involving their mass and so forth, that had not been gotten before.
>
> So this was very successful. And I thought it was interesting, in a way, that the theory used to make the atomic implosion bombs was the same theory I could apply to certain kinds of variable stars. It's interesting to see how things relate to each other. I enjoyed working that out, and I got some distinctions because

although others had tried — there was a group at Los Alamos that had tried to do this using the Los Alamos computer codes — it turned out that it was more effective for me to start from zero and create my own codes... because the bomb codes were so complicated and so hard to use that it was more effective to develop a new code from scratch. And I think there's a lesson there. If you have an all-purpose machine it may not be the best way to do a problem.

Robert demonstrated that the pulsations were driven by the zone in these stars where the second ionization of helium occurs. This is the region where the temperature is roughly 50,000 K (90,000°F), some distance below the star's surface but still far outside its central core. If this region contracts slightly, it gets denser and hotter, and more of the photons streaming from the star's center are absorbed by helium as it becomes doubly ionized. This absorption of energy causes that region of the star to expand outwards, and thereby cool off again. But this cooling causes the region to contract again, and the cycle starts over.

Robert's work predicted how the pulsations changed depending on the color and the brightness of the star, and this prediction was in excellent agreement with the observations. He was even able to predict the detailed peculiarities of the light-curves of these stars as they pulsated. He said,

> I was pursuing it [the Cepheid project] for a number of years, of the order of five years or so. I was trying to improve it, but I ran into problems having to do with convection: it was a physical process that went on in the star, and could not be properly described by the mathematics that I was familiar with — and in fact it still is not described in any great detail, even today. I found I wasn't able to proceed with that, but I was able to carry it far enough that it was rather successful.

Election to the National Academy of Sciences and the Eddington Medal

In 1965, while he was working on variable stars, Robert was elected a member of the National Academy of Sciences. This was a major recognition of Robert's contributions to science, and he received letters congratulating him from a large number of eminent people, as well as from personal friends. For example, J. Robert Oppenheimer and his wife Kitty sent a congratulatory telegram: "Your election good news to all of us. Kitty and Robert." There was a letter from Hubert Humphrey (Vice President of the U.S. under President Johnson), to which Robert replied, "...It is very gratifying that you are so interested in science and scientists. I hope that sometime I can be of some service to you. Yours very

Achieving the Rare

truly, R. F. Christy." There was also a letter from Donald F. Hornig (the White House's Special Assistant for Science and Technology). Alvin F. Weinberg (Director of Oak Ridge National Laboratory) wrote, "…I guess I can only say that I feel much better about being a member of the Academy since it is now apparent that NAS, every so often, recognizes the true excellence you exemplify!" John Archibald Wheeler, the gravitational physicist from Princeton, wrote, "…For me, it will mean more chances than ever to see you — always a pleasure. The problem of pulsations that has occupied you lately is absolutely fascinating." George S.

DR. ROBERT F. CHRISTY
… awarded medal

Medal Given Professor At Caltech

Dr. Robert F. Christy, professor of theoretical physics at Caltech, has been awarded the Eddington Medal of the Royal Astronomical Society of London for h i s calculations on variable stars.

The medal, commemorating the late Sir Arthur Eddington, noted British astrophysicist, is awarded for investigation in astronomy, with preference f o r theoretical astronomy.

Dr. Christy was honored for "his work on the non-linear theory of pulsating stars, which has enabled a close comparison to be made with observations of Cepheids and RR Lyrae variables."

These types of stars are both members of an extensive class whose characteristic variations in brightness are used as basic yardsticks to measure distances within our galaxy and to extend t h e measurements to nearby galaxies.

A member of the Caltech faculty since 1946, Dr. Christy was one of the group that developed the first atomic pile at the University of Chicago in 1942.

At Los Alamos, N.M., he contributed to the design of atomic weapons. It was calculational techniques f i r s t developed at Los Alamos for atomic bomb design that he used in his calculations on variable stars.

He also has worked with Caltech's Kellogg Radiation Laboratory on the theory of nuclear reactions.

News article of February 9, 1967 reproduced courtesy of the Pasadena Star-News

Hammond, a Professor of Chemistry at Caltech, wrote, "It seems to me that this recognition is long overdue and that the action does more to enhance the prestige of the Academy than to honor Christy." One letter even came from a fellow Berkeley student who had not seen Robert for almost three decades, Elizabeth Bond Bickerton.

Two years later, in 1967, Robert was awarded the Eddington Medal of the Royal Astronomical Society of Great Britain for his work on variable stars. This is a medal that is awarded roughly every two years, for major accomplishments in the field of theoretical astrophysics.

Starting to Travel, Late in Life

Early in Robert's life he had no opportunity to travel. It wasn't until he was nearing 50, with his sons at college, that he was finally able to start doing so. He

CEPHEID VARIABLE STARS

R. F. Christy

California Institute of Technology, Pasadena, California

Cepheid-type variable stars have been carefully observed over many years and for many of them, well-determined light curves, which display the variation of light output with time, are known. These curves repeat quite accurately with periods ranging from about 0.1 day at the shortest to 100 days at the longest. Some of these variables, particularly those near 10 days period, show interesting peculiarities, in the form of extra bumps, in their light curves.

Non-linear time-dependent calculations of the dynamics and radiation flow in these stars have made it possible to explore these peculiarities in considerable detail by means of computed models. These computed models show the same kind of peculiarities that are seen observationally and it has been found that in order to reproduce both the observed period and the observed extra bumps in the light curve, it is necessary to fix both the mass and radius of the model. This means that the observation, in two colors, of a particular variable which shows these characteristics leads, by use of the theory, to determination of the mass, radius, absolute luminosity, and distance of the star in question. This technique has recently been applied to determination of the distance of the Magellanic clouds and to determination of the fundamental properties of the Cepheid variables in the clouds. An explanation is proposed to account for the differences between the variables of the Magellanic clouds and those of the galaxy.

Abstract of the talk that Robert presented in 1968 to the National Academy of Sciences usually travelled alone because Dagmar, his first wife, did not enjoy travelling or the outdoors. He had always been extremely drawn to the mountains and in his youth had greatly enjoyed the Rocky Mountains and the Sierras, but had only dreamed of visiting the Alps. Now he was beginning to travel to international conferences in Europe, so he took advantage of this chance to finally experience the Alps.

Robert attended a two-week International Astronomical Union (IAU) meeting in Hamburg in 1964. He used this as an opportunity to buy a Volkswagen in Germany, since it would be cheap there due to a very favorable exchange rate. He drove to Interlaken, Switzerland to go hiking for a day, and then on to Chamonix in France, where there is a famous glacier with fabulous views. He spent a day there before driving through southern France down to the Pyrenees. Driving back north again, he stopped to see the famous Lascaux cave paintings; he was very much impressed by them. He drove north through Paris to Belgium where the car could be shipped back to the U.S. Robert didn't have much money and took considerable pride in the fact that this trip was done on five dollars a day, including gasoline, food, and lodging. Altogether this trip took roughly a month — only rarely did he spend that much time on a trip.

In 1966 Robert attended a variable star meeting in Bamberg, Germany. There he met with German astrophysicists who worked on variable stars and stellar interiors, including Rudolf Kippenhahn and Alfred Weigert. Both of these men would later play a key role in my own life, and in bringing Robert and me together (see Chapter 12).

In 1967 there was an IAU meeting in Alaska associated with the solar eclipse of May 9. Robert flew up to Fairbanks in the center of Alaska. On the way back he took a bus to Dawson in the Yukon Territory of Canada. From there he took a train to Skagway, which is on the southern coastal extension of Alaska. From Skagway he was able to take a boat, travelling along the inner passage to Vancouver — the inner passage is the water route between the mainland and the islands of the Pacific coast. From Vancouver he flew back to Los Angeles.

In August 1970 Robert travelled to the two-week IAU meeting in Brighton in southern England, on the English Channel. From there he flew to Lisbon in Portugal to attend a two-week astrophysics meeting in Ofir (a small town on the coast). At this time he and Dagmar were separated and in the process of obtaining a divorce. He met me in Ofir, and as described in Chapter 12, after only a few days together we both independently decided that we wished to spend our lives together — but there were obstacles to overcome...

A Brief Second Sabbatical, at Cambridge in England

In the spring of 1968, Robert spent four or five months on sabbatical at Cambridge to continue his research on variable stars. As he stated in his 2006 interview with Sara Lippincott:

Robert Christy (Caltech photo)

> Oh, that was fun! I was what was called a Churchill Fellow. This was a fellowship awarded by the fellows of Churchill College, Cambridge, which was a college founded in honor of Winston Churchill, but it had definite connections with the U.S. So they brought over visitors, often from the U.S., to spend time there rubbing shoulders with the students and faculty. In general, it was mutually profitable.
>
> I wanted to go to Cambridge, because I had a sabbatical. I didn't teach there; I did research. It was a wonderful experience. I thought then that Cambridge and Oxford were really the epitome of college experience. I never went so much for Oxford, but Cambridge, I still think, is really tops in terms of what a college should be: the surroundings, the way things are done, everything about it. We don't do things that way in this country. We don't have the tradition.
>
> Now, Harvard has a lot of tradition, but it isn't put together the same way.
>
> I lived there at the college in a small apartment, and I went to seminars and other things having to do with physics and astronomy. It was a most enjoyable experience.
>
> Cambridge had a bunch of good nuclear physicists at the time. And they had radio astronomers. And, of course, in college you meet people outside your own field — the other dons who are living there, or eating there. Dining in college is a very, very attractive experience. I'm all in favor of it. You might say it's very civilized.

While at Cambridge, Robert tried to explore the whereabouts of his paternal ancestors; but he did not get very far in this quest.

On the way back from England Robert spent a week at the first "June Institute" meeting at the University of Toronto, where he was an invited speaker. This is where Robert and I first met (see Chapter 12).

Robert's Last Stellar Pulsation Meeting: Santa Fe in 2009

The field of variable stars in which Robert had been a pioneer has grown immensely since the 1960's. In the first week of June 2009 the latest research results by astrophysicists coming from dozens of countries were presented in one of Robert's favorite places, Santa Fe in New Mexico (close to his beloved Los Alamos).

Even though he was 93 years old and had to travel with an oxygen tank to help his breathing at high altitudes, Robert attended enthusiastically with me and our elder daughter Ilia (who came along because she is a physician and could take care of him if he should have a medical problem).

It was very gratifying for him to listen to all the latest results and to experience how the field had flourished since his pioneering work. He was a guest of honor, treated like a king. Robert presented an animated talk summarizing his life's work in research. He stated that of all the fields of research to which he had contributed, variable stars was his favorite.

At the 2009 stellar pulsation meeting in Santa Fe. Above: Robert is at right with his daughter Ilia. Below: Robert gives a talk at this meeting. (Photos by Thomas E. Beach)

Attendees of the 2009 Stellar Pulsation Meeting in Santa Fe, New Mexico (photo by Kim Simmons):

1. Robert Christy
2. I.-Juliana Sackmann
3. Thebe Medupe
4. Judi Provencal
5. Rafael Garrido
6. Jose Pena
7.
8. Elisabeth Guggenberger
9. Adam Sodor
10. Michael Endl
11. Barbara Castanheira Endl
12. Torben Endl
13. Joanna Molenda-Zakowicz
14. Andrea Kunder
15. Katrien Uytterhoeven
16. Richard Townsend
17. Paul Bradley
18. Katrien Kolenberg
19. Arthur N. Cox
20. Merieme Chadid
21. Stephane Mathis
22. Lucas Macri
23.
24. Ennio Poretti
25. Paolo Battinelli
26. Yoon-sik Jeon
27. Young-Beom Jeon
28. Serge Demers
29. Johanna Jurcsik
30. Lester Fox Machado
31. Agnes Kim
32. Margit Paparo
33. Suklima Guha Niyogi
34. Zsofia Bognar
35. Istvan Dekany
36. Joyce Ann Guzik
37. Savita Mathur
38. Karen Pollard

Achieving the Rare

39. Karen Kinemuchi
40. Michelle Creech-Eakman
41. Catherine Lovekin
42. Charles Kuehn
43. Peter Wood
44. Sasha Kosovichev
45. J. Robert Buchler
46. Claudia Greco
47. Bob Dukes
48. Marcella Marconi
49. Dorota Szczygiel
50. Lee Anne Willson
51. Jennifer Cash
52. Samantha Hoffmann
53. Joy Chavez
54. Pawel Moskalik
55. Mihaly Varadi
56. Marian Doru Suran
57. Maarten Desmet
58. David Kilcrease
59. Luis Balona
60. Patricia Lampens
61. Hiromoto Shibahashi
62. Gulnur Dogan
63. ArneHendon
64. Shashi Kanbur
65. Noriyuki Matsunaga
66. Doug Hoffman
67. Kunegunda Belle
68. David Laney
69. Christine Nicholls
70. Irina Kitiashvili

71. Gunter Houdek
72.
73. Gerald Handler
74. Rafael Garcia
75. Peter De Cat
76. Robert Deupree
77. Wojciech Dziembowski
78. Hilding Neilson
79. Nancy Evans
80. Andrea Miglio
81. Radoslaw Smolec
82. Giuseppe Bono
83. Ann Marie Cody
84. Igor Soszynski
85. Dumitru Pricopi
86. Andrzej Pigulski
87. Don Luttermoser
88. Michael Montgomery
89. Geza Kovacs
90. Chow-Choong Ngeow
91. Martino Romaniello
92. Grzegorz Kopacki
93. John Caldwell
94. James Nemec
95. Matthew Templeton
96. Andrew Jay Onifer
97. Stephen Ratcliff
98. Tom Barnes
99. Edouard Bernard
A.
B. Siobahn Morgan
C. Maxime Spano

D. David Turner
E. Jozsef Benko
F. Joergen Christensen-Dalsgaard
G. Hideyuki Saio
H. Chris Cameron
J. Robert Szabo
K. Jaymie Matthews
L. Vichi Antoci
M.
N. Bruce Hrivnak
O.
P. Steve Becker
Q. Horace Smith
R. Jan Lub
S. David Bersier
T. D. H. McNamara
U. Don Kurtz
V. Ed Schmidt
W. Alexandre Gallenne
X. Thierry Semaan
Y. Alistair Walker
Z. Duncan Wright
AA. Mehdi-Pierre Bouabid
AB. Anne-Laure Huat
AC. Stephane Charpinet
AD. Saskia Hekker
AE. Michel Breger
AF. Travis Metcalfe
AG. Pierre-Olivier Quirion

Chapter 11

As Provost: Achieving His Dreams for Caltech

Robert Christy's first experience of administration took place in 1964 when he became Acting Chairman of the Division of Physics, Mathematics, and Astronomy for two months. In a letter of August 4, 1964 he wrote that this

Robert Christy in his office (Caltech photo)

experience had "reinforced my view that, as of now, I prefer just being a professor of physics." Nonetheless the events of the next few years led him into administration.

Robert was offered administrative posts in other universities, but he declined them all, seldom mentioning them to anyone. One exception to this was in 1966 when Robert was offered the post of Vice Chancellor for Natural Sciences at the University of California at Santa Cruz. He did not dismiss the offer out of hand, but travelled to Santa Cruz with his wife Dagmar to learn more about the position and the area. On the way

back he asked his wife which job he should accept, as he had done previously when he was first considering moving to Caltech from the University of Chicago. As she had done before, Dagmar answered "Caltech." He again followed her advice and stayed at Caltech.

It was on this trip to Santa Cruz that Robert first met Francis Clauser, the Vice-Chancellor at UC Santa Cruz, who shortly thereafter accepted a position at Caltech as Chairman of the Division of Engineering and Applied Science. A deep friendship developed between Robert and Francis Clauser.

Robert Christy (Caltech photo)

169

Robert's involvement in administrative work started shortly after that, when he became head of the search committee to search for a new President for Caltech. Lee DuBridge had been President for more than two decades, from 1946 to 1968. When Richard Nixon was elected President of the U.S. in 1968, Lee DuBridge, a Republican, was nominated as Nixon's Science Advisor. This meant he would be leaving as Caltech's President, and a new one had to be found. In addition some changes in direction were needed, as Caltech's finances were deeply in the red.

Robert's search committee found Harold Brown, who agreed to serve as Caltech's President. He was the second to hold that post. Harold Brown had been the Secretary of the Air Force from 1965 to 1969 under President Johnson. Perhaps he was all the more ready to return to academia after having to deal with the problems from the war in Vietnam.

Executive Officer for Physics and Chairman of the Faculty

In 1968, Robert started a two-year term as Executive Officer for Physics and Chairman of the Faculty at Caltech. He said:

> Chairman of the Faculty is very much a part-time job, in which you are looking after some of the interests of the faculty, but it's not really an administrative job.

So Robert still had time for academics at that point. But he continued:

> I allowed myself to be distracted from research into more administrative activities — in large part because I had come to the end of one piece of research and I did not see yet where the next thing was going to start. My work on variable stars I had pursued as far as I could see how to pursue it, and I had not yet started something new. So it was under these circumstances that I got involved in administration.

Vice President and Provost

Harold Brown picked Robert to be the head of the search committee for Provost. Robert recalled:

> The Provost is the second man in the university, next to the President. And that [the President] was Harold Brown. He picked me to be on a committee to look for a Provost, and I think I was Chairman of that committee. So we selected a

number of very promising names and I traveled around the country to talk to those people and see if I could persuade them to come and take the job of Provost.

I went to interview a number of these candidates, but the candidates were so good that they were needed at the place they were. Several of them later became presidents of the institutions they were at. Bob Sproull of Rochester became president of Rochester. At the University of Maryland, a candidate (I forget his name) later became president of the University of Maryland. There was Harvey Brooks and John Toll... a very distinguished group of people. Thus, although we picked good candidates, we did not persuade them to come to Caltech. In the course of that, I had a number of interactions with Harold Brown, trying to describe what we were doing and our efforts, and as a result after six months or so he decided to ask me to become Provost. I had enjoyed my interaction with him — he was a very capable person, from whom I learned a lot — and I agreed to act as Provost.

I had a very high respect for Harold, because he was a very intelligent person — very well organized in terms of running things, much more so than anyone else I ever knew. And I thought maybe I could help do something for Caltech that way.

The position of Provost at Caltech had first been created in 1961 by then-President Lee DuBridge. Robert's friend Bob Bacher, previously the Chairman of the Division of Physics, Mathematics, and Astronomy, became the first Provost of Caltech at that time. In 1968, two years before Robert accepted the position, President Harold Brown had defined the duties of the Provost:

1) He is the principal adviser to the president on academic matters. This involves the areas of curriculum, divisional structure and organization, research orientation and policy, academic planning including the growth and direction of the institute, and supervision of faculty appointments and policy. This implies that his professional career should have been primarily academic and with some administrative experience.

2) The Provost is the deputy of the President and should be able to substitute for him not only with the institute but also in external relations with the public, with the trustees, and with Washington. This implies that he must be compatible with the president and able to cooperate with him as well as being a good front man in relations outside the Institute.

3) He is the principal officer of the institute in dealing with the faculty. Thus he must understand and be able to represent the faculty view point without being the spokesman of the faculty. He must be able to deal firmly with the faculty and yet retain their respect. He must be dedicated to the Institute.

The job of Provost was a daunting one. In a letter of November 17, 1969 Jack Roberts (who had been Chairman of Chemistry and Chemical Engineering from 1963 to 1968) had written, "I am convinced that it is not really possible for one man ... to carry the load that the faculty, administration, students and trustees seem to expect of the provost." In a letter written a month later he asked *not* to be considered as a candidate for the Provostship.

Below is the news release written by Caltech President Harold Brown describing Robert Christy's appointment as Vice President and Provost:

CALIFORNIA INSTITUTE OF TECHNOLOGY

PASADENA, CALIFORNIA 91109

OFFICE OF THE PRESIDENT   February 19, 1970

N O T I C E

To: The Members of the Faculty

From: Harold Brown *Harold Brown*

After an arduous and diligent search, the Provost Selection Committee has concluded its work, I have recommended and the Board of Trustees has named Professor Robert F. Christy to succeed Professor Bacher as Vice President and Provost of the California Institute. This choice needs no explanation. Judging from my extensive discussions with various segments of the Institute community it will be received with applause.

The official news release follows: Pasadena, Calif.--Dr. Robert F. Christy, professor of theoretical physics and chairman of the faculty of the California Institute of Technology, has been appointed vice president and provost of Caltech, President Harold Brown announced today.

Dr. Christy, 53, will succeed Dr. Robert F. Bacher next fall in this important position, which is the second highest administrative post at Caltech. Dr. Bacher, 64, is retiring after having served as provost since 1962. He will continue as professor of physics.

"As executive officer for physics and as chairman of the faculty, Dr. Christy, who has been a member of the faculty for 24 years, has successfully dealt with problems concerning the faculty and curriculum," Dr. Brown said. "This invaluable experience and the skills with which he has resolved these problems will, I am sure, make him an able successor to Dr. Bacher, who has served Caltech so proficiently for more than a score of years."

As provost and vice president, Dr. Christy will be Caltech's chief academic officer. He will have overall responsibility for faculty appointments and promotions, and for academic planning and research.

Dr. Brown recommended Dr. Christy to the Board of Trustees after extensive consultation with a faculty advisory committee (with student participation) and with senior administrative officers of the Institute.

The Members of the Faculty   -2-   February 19, 1970

"I am looking forward", Dr. Christy said, "to working with President Brown and the division chairmen in helping Caltech continue its fine traditions in science and engineering and in breaking some new ground in the future."

Dr. Christy has been active in faculty affairs for many years, having served on many faculty committees. Among them was the one to select a new president, one on aims and goals, another on academic freedom and tenure and one on academic policies.

As a teacher, he is as demanding as any. However, he believes in arousing the enthusiasm of students to do things on their own, rather than forcing them.

As a scientist, his career has been one of excellence successively in a number of fields. He recently refined important yardsticks for measuring the universe by using a large computer to imitate the pulsations of variable stars. For this work he was awarded the Eddington Medal of the Royal Astronomical Society.

In 1965 he had been elected to the National Academy of Sciences for significant theoretical contributions to several fields of physics, including nuclear, high energy and cosmic ray physics and astrophysics.

He contributed significantly to the theory of neutron diffusion in nuclear reactors, and he was a member of the group at the University of Chicago that developed the first atomic pile.

Born in Vancouver, Canada, Dr. Christy was educated in that city, graduating first in his class from the University of British Columbia in 1935. He obtained his M. S. in physics and mathematics from that university, and a Ph.D. in theoretical physics from the University of California, Berkeley. His professor at Berkeley was the late Dr. J. Robert Oppenheimer, who later invited him to Los Alamos, New Mexico, to work on the first atom bomb during World War II. Dr. Christy was at Los Alamos from 1943 to 1946.

He was an instructor of physics at the Illinois Institute of Technology in 1941-42, and assistant professor at the University of Chicago in 1946. Coming to Caltech as an associate professor in 1946, he was appointed a full professor in 1950.

He is a member of the American Physical Society, International Astronomical Union, American Association for the Advancement of Science, American Astronomical Society, Royal Astronomical Society, and the American Academy of Arts and Sciences.

He is married and the father of two grown sons.

Robert was aware when he took on the job of Provost that it would be a heavy load. However, he also knew that this position would enable him to achieve some of his dreams for Caltech, to fix the problems and make some of the improvements that he had hoped for. He also had some hopes of obtaining a Vice-Provost to help carry some of the load. If it proved possible to create such a position and find someone to fill it, then Robert might actually have some time available for astrophysics.

In a letter to me written in April 1970, a month after he had started in his position as Provost, Robert said:

> So far, I am enjoying much of my new job. I like to get involved in campus wide problems and try to help. I believe, however, that your idea of trying to reserve two months in summer for astrophysics is excellent and I am trying now to lay out this objective and prepare a Vice Provost position to make it possible. You are going to be a great help to me in so many ways!

Article of March 20, 1970 reproduced courtesy of the Pasadena Star-News:

Dr. Christy Provost at Caltech

Takes Second Highest Position

Dr. Robert F. Christy, professor of theoretical p h y s i c s and chairman of the faculty of the California Institute of Technology has been appointed vice president and provost of Caltech, President Harold Brown announced Thursday.

Dr. Christy, 53, will succeed Dr. Robert F. Bacher next fall in this i m p o r t a n t position, which is the second highest administrative post at Caltech. Dr. Bacher, 64, is retiring after having served as provost since 1962. He will continue as professor of physics.

"As executive officer for physics and as chairman of the faculty, Dr. Christy, who has been a member of the faculty for 24 years, has successfully dealt with problems concerning the faculty and curriculum," Dr. Brown said.

As provost and vice·president, Dr. C h r i s t y will be Caltech's chief academic officer. He will have overall responsibility for faculty appointments and promotions and for academic planning and research.

Won Award

He recently refined impor-tant yardsticks for measuring the universe by using a large computer to imitate the pulsations of variable stars. For this work he was awarded the Eddington Medal of the Royal Astronomical Society.

In 1965 he was elected to the National Academy of Sciences for significant theoretical contributions to several fields of physics, including nuclear, high energy and cosmic ray physics and astrophysics.

He contributed significantly to the theory of neutron diffusion in nuclear reactors, and he was a member of the group at the University of Chicago that developed the first atomic pile.

Canadian Born

Born in Vancouver, B.C., Canada, Dr. Christy was educated in that city, graduating first in his class from the University of British Colombia in 1935. He obtained his M.S. in physics and mathematics from that university, and a Ph.D. in theoretical physics from the University of California, Berkeley. His professor at Berkeley was the late Dr. J. Robert Oppenheimer, who later invited him to Los Alamos, N.M., to work on the first atom bomb during World War II. Dr. Christy was at Los Alamos from 1943 to 1946.

He was an instructor of physics at the Illinois Institute of Technology in 1941-42, and assistant professor at the University of Chicago in 1946. Coming to Caltech as an associate professor in 1946, he was appointed a full professor in 1950.

He is married and the father of two grown sons.

Article from the March 1970 issue of the Caltech News:

Physicist Robert F. Christy

Caltech Names New Provost

Caltech will have a new vice president and provost next fall, and even though he has been a member of its faculty for 24 years, he looks forward to having six months to learn about the job before he has to start doing it. Robert F. Christy, professor of theoretical physics and chairman of the faculty, will hold the second highest administrative post at the Institute when he succeeds Robert F. Bacher, who retires after having served as provost since 1962.

As Caltech's chief academic officer, Christy will have over-all responsibility for faculty appointments and promotions, and for academic planning and research. He feels that the fact that he has been at Caltech for a long time, that he knows the faculty, and that they know and trust him will be very helpful.

Christy decided to accept the job of provost partly on the basis that he has changed the nature of his work occasionally in the past and has found doing something new very stimulating. "I'd rather try new things than get into a rut," he says.

In fact, Christy's career shows very little evidence of time spent in a rut. During World War II he worked at Los Alamos on the development of the atom bomb. In the immediate postwar years he concentrated on theoretical and nuclear physics, and for the last nine years he

Continued on page 3

Christy: Some Interesting Changes

Continued from page 1

has been doing astrophysics, with the result that his calculations on variable stars won him the prestigious Eddington Medal of the Royal Society of London in 1967.

His research and teaching contributions are only part of his service to Caltech. He has been on the faculty board, the academic policies and the academic freedom and tenure committees; he was also a member of the presidential selection and the aims and goals committees. He became executive officer for physics in 1968, and in 1969 was elected chairman of the faculty. He says he spent a

good deal of time in that newest job interacting with faculty members, discussing their problems, and trying to see sensible approaches to helping them. It was an interesting experience, and it led him to believe that what a provost would have to deal with would be both challenging and rewarding.

The time Christy has spent on the chairman's job in the last six months has interfered considerably with his research, so he is prepared for the fact that being provost will interfere even more. What makes it worthwhile, however, is the opportunity to oversee the activities he feels are the very heart of the Institute—its academic program.. (He stresses that he means "oversee," not supervise or control.)

Because the provost is supposed to be aware of and sensitive to faculty views, reactions, and problems, and because the provost has over-all responsibility for academic matters, Christy believes that in the long run the actions of the provost will have a major cumulative effect on the Institute. However, Caltech's small size and the accessibility of its administrative officers to anyone who wants to speak to them make all activities and problems somewhat interrelated.

"At this juncture in the Institute's history," Christy comments, "there are some very interesting changes in the works. In an administration where many positions are filled by persons who have held them for a long time, a newcomer to a job is constrained by traditional habits and attitudes. We have a new president and many other new people in important positions or being sought for them. We are likely to be bound much less than normally by the past. This is a situation that invites expression of new ideas and gives freedom for new ways of doing things."

Broadening Caltech's Focus: Humanities and Economics

There was some debate about just how involved a technical institute like Caltech should get in teaching humanities and economics. Robert recalled:

My concerns for Caltech were to try to keep it as good as it was. Under Millikan and DuBridge, it had developed into a fantastic institution, and mostly I wanted to keep it at that level. Harold wanted to broaden it a little. He was the one who acquiesced in the matter of bringing the social sciences here.

We already had a humanities division, and it was always a little bit restless, because its members were, you might say, second class citizens in a certain way. They were always highly respected — intelligent people who were excellent teachers — but it was generally recognized that they had not been chosen primarily as scholars but because they were good teachers. And Caltech was very lucky to have them. They gave the students an excellent education. But they were not viewed in the same way as the scientists, and they had aspirations to be treated more equally, in terms of what they were permitted to do. They had aspirations to broaden the division to include the social sciences. So that did take place, and Harold supported it — I might say, with rather more enthusiasm than I did.

The feelings among the faculty were mixed. Some probably felt that it was good for Caltech, and others mistrusted it.

Economics was brought in. There wasn't much psychology, though there was a little. We did have a kind of psychology that tied into biology — psychobiology. That was started here in a rather respectable way by Roger Sperry, who did split-brain research. But it never tied into the humanities: it was the biological part of psychobiology that worked out. But the hope at that time was to find some kind of psychology that would bridge the humanities and biology. I don't think that ever worked.

Economics was emphasized. At least there was a kind of precise competence involved in economics. I never opposed it, because my boss was helping that development, and I worked with my boss. But I never had the same enthusiasm for it.

Bringing Caltech Back into the Black Financially

In 1969 there was a problem at Caltech in that during the DuBridge administration Caltech had been fallen seriously into the red financially. Robert Gilmore, the Vice-President for Financial Affairs, had made poor financial decisions, and so had others in the administration. This fiscal problem was surprising in that the post-war years were golden years financially for science in general, with a great deal of government support. Science was considered well worth investing in, partly because it was recognized that scientific breakthroughs such as radar and the atom bomb had been crucial to ending World War II.

David Morrisroe, a consultant for general management problems with an MBA from Harvard, was brought to Caltech in 1969 as part of a study by outside advisers to suggest better means of managing Caltech's endowment. At this time Robert Christy was working with Provost Bob Bacher to learn administration before becoming Provost himself. (This "apprenticeship" for the position of Provost was the idea of Bacher, a very skilled and effective administrator. Going from a professorship to a powerful administrative position is a huge leap, and Bacher wanted to make sure that Robert would be well prepared. Robert does not recall hearing of any such "apprenticeship" since that time. People just learned from their mistakes on the job — sometimes with serious consequences.)

Robert was very impressed by David Morrisroe and pushed to have him offered a position at Caltech. Robert took the unusual step of getting Robert Gilmore removed from his responsibility for Caltech's investments, replacing him with Morrisroe, who became Caltech's Director of Financial Services in 1969. Five years later, after Morrisroe had proved himself to be unusually capable, he was promoted to Vice-President for Financial Affairs and Treasurer (Gilmore's previous position). Morrisroe served Caltech for over a quarter of a century under four different Presidents, and his expertise was crucial in stabilizing Caltech's finances. Robert's selection of Morrisroe turned out to be a stroke of genius, with tremendous impact on Caltech's well-being.

The CALIFORNIA Tech

Volume LXXVI   Pasadena, California, Friday, April 25, 1975

Ivory Tower under Siege

Caltech Depression

by Richard Gruner

"The late 1960s and the early 1970s brought—along with ferment and controversy— financial stresses that threatened most American universities, including Caltech ... We responded to these various financial problems by putting further controls on spending ... As a result, Caltech has weathered the financial storms of 1969-1972 better than many other universities ..."

Harold Brown
Caltech President
January 1974

"There's no cushion left. We're right at the edge of an economic morass."

Robert Christy
Caltech Provost
April, 1975

During the fifteen months between the statements above, Caltech's finances have been severely battered. Increased operating costs coupled with reduced endowment income and altered giving patterns have greatly weakened the Institute's economic underpinning. As a result, numerous cuts in Caltech support service expenditures were necessary during the past year and when Provost Robert Christy announced early this month that the Institute was delaying new faculty appointments for lack of funds, it was clear that its present economic crisis might seriously jeopardize Caltech's position "at the leading edge" of science.

Caltech's current financial difficulties are closely tied to the general depression in the U.S. economy. Stock market setbacks in particular, have significantly reduced the Institute's income from its endowments. Long noted for their investment skill, Caltech economic planners find some consolation in the fact that the value of Institute stocks wend down only 22% during a period when the Standard and Poor stock index dropped 50%. There is little doubt, however, that recent losses in the stock market have created significant financial problems for the Institute.

Unrestricted giving is another important source of income which has decreased during the current U.S. depression. Although the total amount of gifts received by the Institute has remained relatively constant (in contrast to significant decreases at many other U.S. universities), more and more donors are earmarking gifts for specific purposes such as endowed chairs. While these restricted gifts sometimes subsidize activities which previously required general funds, overall they have a much less important effect upon general fund revenues than do their unrestricted counterparts. Thus the newly altered giving patterns of Institute donors have greatly aggravated Caltech's general fund woes.

Utilities Increase

Caltech's income crisis has come at a time when many costs are on the rise. One of the largest (and least anticipated) areas of increase was utilities expenditures. Between 1971 and 1976 utilities added $0.8 million dollars to the Institute budget. In other areas cost increases have been less spectacular, but the total increase in the Institute's financial burden has been substantial during recent years. This increased burden coupled with decreased general fund income has forced the administration to conclude that "in order to live in Fiscal 1977 as we have in the past, we need one million dollars added to our

Continued on Page Six

RECESSION AT CALTECH. Provost Christy predicts financial hardships for Caltech in the years ahead.

Continued from Page One

general funds budget" according to Provost Robert Christy.

Significantly, Caltech administrators expect academic operations to absorb a large share of the expenditure cuts necessary to erase the Institute's projected one million dollar deficit. Over the past five years, general budget increases of 4.1% per year have resulted from academic operations increases of 5.8% and supporting services increases of 2.9%. The low supporting services growth rate has been achieved through numerous expenditure reductions and has left supporting services able to absorb few further cuts. Thus, during Caltech's present financial difficulties the Institute administration has turned to academic operations with an eye towards cutbacks.

In all, according to Dr. Christy, academic operations are expected to absorb a 0.3 million dollar drop from last year's level of Institute support. Of this amount, it is anticipated that over half will be replaced by new external research support. The rest, about 0.1 million, will be saved through delaying new faculty appointments.

Faculty Promotions

The present make-up of Caltech's faculty indicates one reason why such an action is significant. As can be seen from graphs (1) and (2), 1969 marked the end of the steady rate of faculty growth which

Caltech had enjoyed since the thirties. As graph (3) shows, although the overall size of the faculty began stabilizing in 1969, faculty advancements and grants of tenure continued at high rates until 1971. According to Dr. Christy, this may have occurred because of his unfamiliarity with the job of Provost (he was appointed in 1970) since "it takes an experienced Provost to stand in the way of Faculty promotions." Whatever the cause, the period 1969 to 1971 left the Institute extremely "top heavy" with 83% of the faculty tenured and 70% holding full professorships.

These high percentages are disadvantageous for two reasons. First, the large financial commitment necessary to support a faculty so highly tenured and consisting of almost 3/4 full personnel could be disastrous.

Deficit Spending

If delaying appointments is undesirable, what then, are the alternatives? One course proposed by economics professor Roger Noll is that the Institute should be more optimistic about future financial developments and adopt a temporary deficit spending policy. According to Noll, Caltech's use of three-year averages of past economic conditions to predict future income levels has given too much weight to downward trends last year. Consequently, Noll recommends that the Institute make an increased draw upon its endowment with confidence that recent financial losses merely represent the bottom of a cyclical variation.

Although he agrees that a deficit spending approach may seem helpful in the short run, Caltech Vice President for Financial Services and Treasurer David Morrisroe feels that additional draw upon the Caltech endowment would lead the Institute into a financial trap of rapidly increasing endowment encroachments. According to Morrisroe, this has been the experience of other mejor private institutions such as Harvard, Stanford and the University of Chicago.

Salary Freeze

Another possible alternative to delaying new faculty appointments would be a postponement of faculty salary increases. This possibility was seriously suggested at a recent faculty board

meeting and Provost Christy feels that a faculty referendum upon this question might approve delayed salary hikes. Despite this, Christy asserts that Caltech as a whole might suffer by this course of action if it resulted in certain leading professors being "bought away" from the Institute by other schools.

· Assessing the disadvantages of each of these alternatives, the administration has found delaying new appointments the least distasteful. Conservative in its approach towards the Institute's endowment and future salary commitments associated with a highly tenured faculty and wishing to avoid the sensitive question of faculty pay cuts, administration leaders have set upon a course which may hope to avoid the worst dangers.

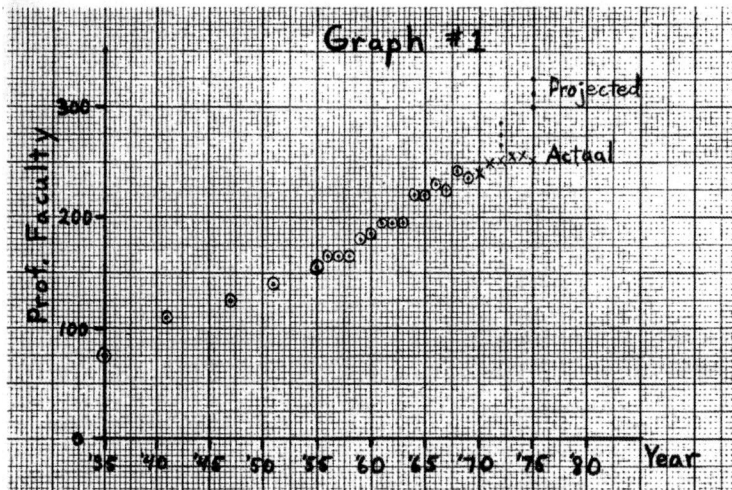

GRAPH ONE:Professional Faculty versus Year(Actual).

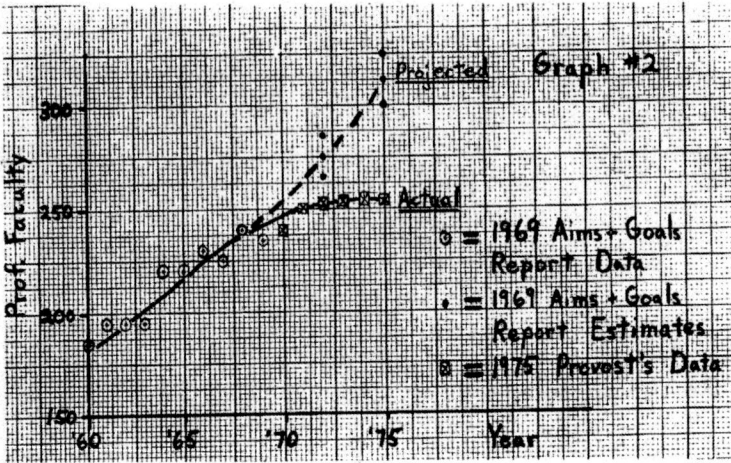

GRAPH TWO:Professional Faculty versus Year(projected and actual).

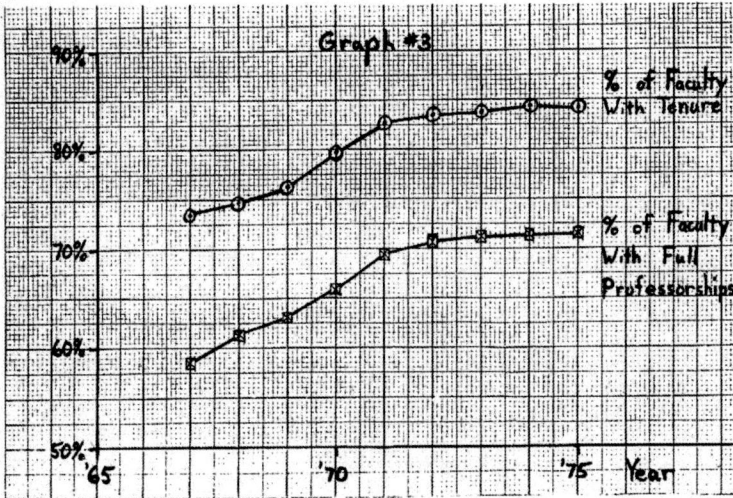

GRAPH THREE:Percent of Faculty with Tenure(and with Full Professorships) versus Year.

President Harold Brown and Provost Robert Christy worked hand in hand on controlling the outflow of money for administrative expenses. Harold and his wife Colene led by example.

When the beautiful and historic administrative building, Throop Hall, was damaged in the earthquake of February 1971 it would have been very costly to repair it and bring it up to modern earthquake standards. To preserve the funds for faculty and students, and also to provide an example to the rest of Los Angeles, Harold and Robert decided to demolish Throop Hall. Nor did they spend money on a new administrative building. Instead they moved into modest quarters on the third floor of the Millikan Library, the main library at Caltech.

When the Browns did major entertaining at their residence for the Trustees and distinguished visitors, Colene Brown would save Caltech's funds by making the flower arrangements herself, with the help of her secretary. She would sometimes even prepare the delicious dinners herself. When Robert and I entertained (due to Robert's position as Vice-President and Provost) we followed in the Browns' footsteps, to give a personal touch and save money for Caltech.

Above and below: Robert Christy serving Caltech as its chief academic administrator (Caltech photos)

I remember that when Trustees would come to stay overnight, Colene would go out of her way to bring a personal bouquet of flowers to the hotel rooms where these guests were staying. Sometimes Colene even had distinguished visitors, including Trustees, stay at the Browns' own house — it could require a major effort to prepare their home for such a visit.

Robert declined to fly first-class at Caltech's expense. He felt that only the President, who needed to represent Caltech and make contacts, needed to fly first-class. Robert also insisted on keeping his office staff small — only two aides — in order to save funds for the real purpose of Caltech, namely its faculty and students. He chose to do as much work as possible himself, giving up the research which was his real love and foregoing all the recognition one receives from scientific accomplishments. This is in contrast to many others in his position who hired large staffs on whom to offload administrative duties while keeping their research going. This is another example of Robert's generosity. He even gave up his office in the Physics building to make space for young faculty, and experienced great difficulty getting a suitable office back later when Goldberger became Caltech's President and hired a new Provost.

New Young Faculty and Productive Older Faculty

Both Bob Bacher and Robert Christy felt that the faculty was the lifeblood of a university, and that the very best must be brought to Caltech. However, faculty requires not only funding but also office and laboratory space. All of these were limited and would be very difficult to expand when Caltech's finances were in the red. Also Caltech did not wish to expand into a huge institution. One of Robert's innovative techniques was to encourage early retirement at ages of 62 or 65 so that office and laboratory space would be available for young incoming faculty. Robert worked out a method whereby professors could retire at 65 and have a similar income as before, since Social Security would pay part of the retirees' income — and that did not come out of Caltech's funds. The funds thus saved became available for new faculty. This allowed Caltech to hire brilliant young faculty who continued to keep Caltech one of the prime science centers of the U.S. and, indeed, of the world.

On the other hand, Robert was also willing to make exceptions to allow productive older professors to keep their laboratory space and students beyond the age of 70, which was the forced retirement age. One prominent example of this was Max Delbrück, who had received the Nobel Prize for Physiology/ Medicine in 1969. Robert persuaded the Trustees to create a new position and title "Board of Trustees Professor" for the oldest professors who were still very productive, so that they could continue rather than being forced into retirement.

The Jenijoy La Belle Court Case

In 1969 Jenijoy La Belle became the first woman hired as an assistant professor at Caltech in literature. In 1974 she was unanimously recommended for tenure by the department faculty. However, the Chairman of the Humanities and Social Sciences Division, Bob Huttenback, informed her that her tenure was denied due to lack of "scholarly productivity" — even though two men with fewer publications than La Belle had just recently gotten tenure. The literature faculty protested the decision. An outside arbitrator was brought in and La Belle agreed to follow his verdict. She assumed that the Chairman had made the same agreement, but when the arbitrator ruled in favor of La Belle the Chairman denied any such agreement and once again refused La Belle's tenure.

The Chairman then proceeded to change the criteria for tenure to include acceptance of a book for publication by a major university press. In a matter of months, La Belle's book was accepted by Princeton University Press. She asked for tenure reconsideration. Again it was denied.

La Belle filed a complaint with the Equal Employment Opportunity Commission (EEOC). Due to this court case, Robert's office was bombarded with endless requests for documentation. Basically, the administration was left with no time to carry out any of its other work. Robert looked into the case, and together with Caltech Trustee Lew Wasserman, pushed for what they considered to be right: settling the case and granting tenure to Jenijoy La Belle as soon as possible (in contrast to the Division Chairman's recommendation). Robert eventually managed to persuade those who opposed this to go along with his decision. It was a great relief when this case was over and the administration could go back to actually running Caltech.

In 1977 the EEOC ruled in La Belle's favor, charging Caltech with discrimination against women faculty. The EEOC also discovered that La Belle had been paid less than her male colleagues of equal rank. Armed with these findings, La Belle met with university trustees. They negotiated a settlement: assistant professorship and promise of an unbiased tenure review in two years. In April 1979 Jenijoy La Belle became one of the first tenured women on the Caltech faculty.

Since then Jenijoy La Belle has been a popular and respected professor at Caltech.

Classified Research at the Jet Propulsion Laboratory

The Jet Propulsion Laboratory (JPL) in Pasadena five miles from Caltech's campus is part of the National Aeronautics and Space Administration (NASA). It was commissioned to be the main center in the U.S. for unmanned space exploration, and receives a significant amount of funding from NASA. On the other hand, JPL also has a loose relationship with Caltech, both via administration and because many Caltech graduates and professors obtain positions at JPL. Caltech receives a significant fee for being involved in the administration of JPL.

On the other hand, unlike Caltech, JPL is a government agency and there is classified work going on there. Robert always felt that successful research required free and open communication. This already had been a problem at Los Alamos, where Oppenheimer had had to fight to maintain free communication even *within* the Manhattan Project. Robert said:

> There was a problem in that Caltech, the essence of a scientific institution, is to find new knowledge and communicate it. That principle is antithetical to classification and keeping things secret. So, in general, universities have long tried to stay clear of classified work, because it's not the way universities should run...
>
> Well, JPL was connected to Caltech. And we did not want JPL to be constrained by secrecy to the point where people there could not talk to people on the campus. And yet some of the jobs JPL undertook had to have some secrecy. So there was a real problem.
>
> Most of JPL's work was not military work. But there was a time when NASA's budget was getting very tight, and JPL, being a large organization, had a large number of people that they had to keep employed. And for a while they found it necessary to take in some classified work for the Department of Defense. But we did not want this to change the character of JPL. We wanted to keep the military work as a restricted thing: it was not going to be the general pattern of JPL.

Robert Christy served on the Advisory Council for JPL in the 1960's. He was among those involved in choosing Directors of JPL. William Pickering had been the illustrious first Director of JPL, a very dynamic person who shaped the young institution. When he retired in 1976 it was difficult to find someone to fill his shoes. Robert was serving as Vice-President of Academic Affairs and Provost, and shortly thereafter the Interim President of Caltech, and was active in finding a promising new head for JPL. This was Bruce Murray, a Caltech professor of planetary science.

In 1982 Robert was also active in selecting the third Director of JPL, Lew Allen, Jr., who had been Chief of Staff of the Air Force.

Opposing the Strategic Defense Initiative ("Star Wars")

The SDI project proposed to put nuclear warheads in orbit around the earth which would power single-use X-ray lasers when they exploded. These lasers were supposed to be used to shoot down incoming ballistic missiles, assumed to be coming from the Soviet Union. This system was intended to act as a "protective shield" above the U.S. Robert Christy, Hans Bethe, and most of the giants who had created the first nuclear bomb were strongly opposed to putting nuclear bombs in orbit above the earth. In addition, most scientists who had studied this project were convinced that its goals could not be achieved. Any such "shield" would let far too many missiles through, and would merely encourage nuclear proliferation without providing any real protection.

In the mid-1980's Robert was the Chairman of the Committee on Oversight of Classified Research at the Jet Propulsion Laboratory (JPL), which has always been closely affiliated with Caltech. In 1984 the issue arose about SDI work taking place at JPL, and this was discussed in a semi-public meeting at Caltech. Robert focused on the issue of weapons research. The most difficult question, he considered, was the "widespread and deep mistrust of the SDI within the Caltech community." However, Robert said that he saw a chance of proving that the goals of SDI were "illusory." Robert believed that SDI was not technologically feasible.

Edward Teller, the father of the hydrogen bomb, had been pushing SDI at the highest levels of the U.S. government. However, in 1987 evidence appeared that he might have been providing overly optimistic estimates of the technical feasibility of this project. Some informed scientists went so far as to say that Teller had been presenting fraudulent evidence. Robert, who had known Teller as a colleague during the Manhattan Project years and in 1946 had shared a house with him for six months in Chicago, commented that, "He [Teller] has a very, very strong fear of the Soviet Union. A lot of his actions are governed by that."

Student Unrest at Caltech

In the 1960's and 1970's, when there was a great deal of student unrest in the U.S., some such incidents occurred at Caltech. At one time when Robert was Provost students surrounded Throop Hall, the administrative headquarters for the President and Provost of Caltech at that time. Rather than trying to ignore them or sending Caltech staff or security personnel out to face the protesting students, Robert went out himself to stand up in front of the mob. With his calm personality, he persuaded them to back down from their demand. In his words,

> They wanted to lower the flag for some reason that was not part of Caltech's general ritual for lowering the flag. And I defended the flag. But Caltech

students were never very far out. They were a very good bunch of students, but they had a few wild times. I think most of the unrest then was on the part of the humanities students rather than the science types, because the science types were too busy working. That's probably why Caltech came through with very little trouble — simply because people worked hard with their noses to the grindstone. They had very definite programs in mind. They didn't have time to let their thoughts wander all over as to what's going on in the world.

The First Women Undergraduate Students

There were some woman graduate students at Caltech a good deal earlier than undergraduates. The first woman to receive a Ph.D. from Caltech was Dorothy Ann Semenov, who obtained her degree in chemistry and biology in 1955. The first woman faculty at Caltech was the mathematician Olga Taussky Todd, who became a Research Associate in 1957 (at the same time as her husband was named professor of mathematics at Caltec). She was the only woman faculty member until 1969.

Photo and caption of February 27, 1976 reproduced courtesy of the Pasadena Star-News:

WOMAN OF THE YEAR — Dr. E. Margaret Burbidge, center, professor of astronomy at UC San Diego, received the "Woman of the Year" award for her outstanding research from the Muses, women's support group of California Museum of Science and Industry, Los Angeles. She is pictured with Dr. Juliane Christy, left, Caltech astronomer, and Mrs. Harold Brown, wife of Caltech's president.

In 1968 Caltech's President Lee DuBridge began to prepare for woman undergraduates by building a coed residence. A great deal of thought had to be put into how to make them feel welcome and to create facilities for them (for example, adding women's locker rooms at the Caltech gym). In the fall of 1970 the first women undergraduates were finally admitted, with strong support from Caltech's new President Harold Brown and new Provost Robert Christy.

Women Faculty at Caltech, and Elimination of Caltech's Nepotism Rule

I recall an incident at a conference on Stellar Evolution and Variable Stars in Ofir, Portugal in September of 1970. At that conference Robert Christy tended to eat lunch with the leading administrators of astronomical institutions. Robert and I had become romantically interested in each other, and one day he escorted me on his arm to his table to join his lunch group. Present were Richard van der Riet Woolley, the Astronomer Royal of England; Olin J. Eggen, the Director of the Mount Stromlo Observatory in Australia; and three leading administrators from other countries. Each of these had hundreds of people working at their respective institutions.

As Robert and I approached the table, Eggen pointedly remarked that marriages should be abolished. (Presumably he knew that Robert had been married, but possibly not that he and his wife had separated — and that he now was interested in me.) All five of these leading male scientists were reciting how, for their respective institutions, there were no women to be found who were capable enough to be hired. Christy agreed, and stated that in addition nepotism rules existed at Caltech, such that a husband and wife could not be hired at the same institution. Caltech was a small institution and it was feared that a husband and wife would always vote the same on committees, and thus would have undue influence.

Eggen stated that only three hundred of the three thousand astronomers in the world were real professionals, and that Professor Helen Sawyer Hogg of the University of Toronto could be considered no more than an amateur (and her husband too). She happened to one of my role models, and I felt personally attacked. They continued with their derogatory remarks about woman professionals, and Woolley went so far as to say that a woman was best suited to be a "pretty ornament" in the home of her husband. Robert Christy listened quietly and did not contradict them. I looked pointedly at each of them, one by one, took my purse, got up, and left their table, saying, "Gentlemen, you don't deserve my company."

Later, privately, I questioned Robert Christy about the case where a father and son (Charlie and Tommy Lauritsen) had both been hired as tenured physics professors in the same department at Caltech. This was in the Kellogg Radiation Laboratory where Robert himself had worked. I also questioned him about Olga Taussky Todd and her husband, who were both mathematicians on the Caltech faculty. He told me that these were exceptions. I asked him whether they had

undue influence on the committees and he said "no." I asked him if he as
Provost (Academic Vice-President) had the power to change the nepotism rule.
He said "yes." I said, "Well, then change it, or you will have to come and get a
position in Germany where these rules don't exist, if you want to be with me."

On October 26, 1970, about a month after Robert had returned to Caltech, he
wrote to me:

> I have recently, among other activities, been formulating a new Caltech policy on
> employment of close relatives of faculty — you recall our discussion at Ofir? It
> permits employment of anyone with any relation but prohibits use of
> recommendations from a close relative. I have become much more sensitive to
> this problem recently for reasons you understand only too well. I think the new
> proposals are defensible on all sides.

When Willy Fowler first hired me as a postdoctoral research fellow at Caltech
in 1971 the nepotism rule was not relevant, as I was unmarried. By the time
Robert and I married in August 1973, Caltech's administration (i.e., Robert) had
eliminated the nepotism rule.

At Nixon's "Western White House"

In the spring of 1971 Robert was among those invited by President Richard
Nixon to the "Western White House" (Nixon's estate in San Clemente) to discuss
science issues of national importance. In a letter to me dated March 31, 1971,
Robert wrote:

> Tomorrow morning I have been invited to a meeting at the Western White House.
> It is situated at San Clemente on the beach about 20 miles south of Laguna where
> we were when you visited. I will leave early as the meeting is for 9:30 and
> continues through lunch. The subject has to do with unemployment in the
> aerospace industry and I will have to see whether I can make any contributions.

In another letter dated April 3, 1971, Robert continued:

> Well on Thursday I did visit the Western White House and attended an all
> morning meeting. I left at 7:30 AM and drove South via the Santa Ana freeway
> as we did going to Laguna. But I went on further south past San Juan Capistrano
> to where the freeway meets the ocean at San Clemente. There I followed
> directions and ended up at a guarded gate leading to the Presidential Compound.
> I parked at an administration building — a temporary structure. The grounds
> were not particularly attractive where I was — parking lots, helicopter pad, and
> temporary buildings. Fog prevented me from seeing if there was a prettier part.

I was directed into a conference room about 15 × 30 feet and joined some others who had arrived early. Eventually there were a few big industrialists — president of General Motors, President of Southern Pacific Railroad, some aerospace executives, some representatives of engineering societies, Provosts from Stanford and Caltech (in place of their presidents) and the President's Science Advisor (Mr. David), the Secretary of Labor (Hodgson) and for one hour President Nixon.

So much for the set up. Everyone except me made comments on the unemployment problem among scientists and engineers. The President spoke for an hour. When he entered the room he was followed by about 50 photographers who shot for 5 minutes and then left. Since I was sitting just one removed from the President (and the cameras were aimed at him) I appeared later on national television (still saying nothing).

The whole thing was rather a farce since it was planned as a publicity vehicle for the President to have him appear to be doing something about the unemployment problem. I presume it was successful in that respect. As a conference I don't think it accomplished anything. But it was interesting to be there and see the President in action at close range. But I would not have wanted to go to Washington just for that.

So it was fortunate that Robert had only had to travel as far as San Clemente…

At Nixon's "Western White House." In the back row, Robert Christy is at left, and President Nixon is the third from the left. (Caltech photo)

Advising on Science in Israel and Egypt

In 1973 Robert was invited to go to Tel Aviv in Israel to provide advice on how to organize science education.

In 1975 Robert was invited by the President Anwar Sadat of Egypt to advise Egypt on the formation of national scientific institutions. Robert left for Cairo, but I reluctantly stayed behind to take care of our 8-month-old daughter, Ilia. (Recall that Anwar Sadat was the President of Egypt who in the October War of 1973 had recovered the Egyptian territory lost in the Six Day War of 1967, but who then negotiated the Egypt-Israel Peace Treaty, for which he was awarded the Nobel Peace Prize — though this also led to his assassination in 1981.)

Hazards Associated with Nuclear Energy

Robert had become a member of the prestigious National Academy of Sciences in 1965. He was a member of several committees of the Academy at various points in time.

In 1975, the National Academy of Sciences decided to set up a committee to review the risks associated with nuclear power programs. Robert Christy was chosen as the Chairman of this nine-member committee: he organized the committee and developed its working procedures. In 1976 he resigned from the chairmanship due to the heavy duties he was beginning to take on when he became Interim President as well as Provost of Caltech, but he continued to work on the project as a member of the committee. The result was a 1979 report entitled "Risks Associated with Nuclear Power." Robert recalled:

> I was also on other committees of the academy. I remember one committee, that was to review the hazards of nuclear power. We tried to examine thoroughly all of the problems and hazards that could arise, starting from the mining of uranium ore, to the operation of reactors and the commissioning of reactors. We examined many questions and spent a long time, with various meetings in Washington and elsewhere, to review these questions and put out a long report. This was at a time when there was considerable environmental concern about nuclear energy, and we were trying to make a thorough review of what the possible hazards might be; and that report exists.

Robert giving presentations while in Tel Aviv, Israel in 1973

We evaluated the hazards. There was no real "result," except that the report would be a useful thing for someone who was interested in questions of nuclear energy, to review what it was we considered. I mean, you could examine the hazards of automobiles, and the hazards are numerous, from the manufacture of steel all the way through to the digging of oil wells and the operation of refineries to the making of rubber, the operation of vehicles on the road, and building of roads. There are hazards associated with all of these things, and which thus are associated with the use of automobiles. We tried to do that kind of analysis for nuclear energy. I don't remember exactly when this was done; I would guess probably in the sixties sometime, or the early seventies.

Chapter 12

The Second Marriage, to I.-Juliana Sackmann

Background of Inge-Juliana Sackmann

I was born in Germany on February 8, 1942, in the middle of the Second World War.

By January 1945 when I was nearly three years old the advancing Russian front had come close to our beautiful country estate in Prussia, and we could hear the artillery fire. The Nazis had placed officers in our home both to watch us and to guard my family's business (large flour mills). The Nazis had ordered everyone to stay in place and resist the Russian invasion.

On the night of January 19, 1945 my family secretly prepared, with the help of our loyal Polish employees, to send the women and children westward. Cars could not be used as gasoline was unavailable. The rare trains were reserved for military uses. In the middle of the night five of our horses were hitched to three large covered wagons that had been used to transport supplies. My grandfather Gottfried Stelter, his partners, and their male employees were to stay behind as cover for the flight of the women and children. My grandmother Juliana Stelter, my mother Lilly Sackmann, and the wife of one of the partners drove the wagon teams while the toddlers rode in the wagons, wrapped in warm clothing and bedding against the bitter cold. The only man with us was my father, Emil Sackmann. He had been lying wounded in a nearby military hospital and was given permission to come home for a single night to celebrate the first birthday of his youngest child Ute, but he was not in good enough health to drive a team of horses. The wagons carried a large amount of food for both the people and the horses, as well as extra warm clothing and precious personal belongings. These last consisted of personal photographs, small oriental rugs, fine china, and sterling silver.

Since then horses have always remained a symbol of freedom and joy to me, and I made them a big part of my later life.

We first travelled to Berlin in central Germany, but did not find safety there. The air strikes by the Allies (always referred to as "the Amis", a German derogatory term for the Americans) were coming down brutally, collapsing the beautiful homes and buildings around us. We would scurry into the basement for fear of being crushed in such a collapse. I also remember fleeing into the subway tunnels until they were flooded by the Nazis. I remember hearing at the time that this had been ordered by Hitler to punish the refugees for leaving their homes and for failing to resist the invading enemy. Later my family fled to Dresden. We hoped that this city would be free of any military presence, and therefore not a target for bombing raids. We were just outside Dresden when the firestorm ignited by the Allied air strikes destroyed this ancient and beautiful city.

Due to the stress of the war and the post-war years, my father was looking for a new beginning. In 1947 he insisted on a divorce from my mother. She loved him and could not agree to a divorce, resisting it for three years. To try to persuade him to stay with her and to demonstrate that she still trusted him, she signed all her property over to him. In 1950, when I was eight years old, she finally had to give up and he got the divorce that he wanted. His lawyers saw to it that she received no alimony and almost no money for child-care expenses. The court gave custody of myself and my middle sister Christel to my father, while custody of my youngest sister Ute was given to my mother. I pleaded with my father to be allowed to stay with my mother. He allowed me to stay with her during the week but would come to pick me up on weekends. I lived in fear of the day when he would not allow me to return to my mother. This fear and uncertainty had a very negative effect on my performance at school.

My mother and I were extremely upset about my father, his lawyers, and the German court system. My mother took any paid job that she could find. This was the first time that she had needed a job. She had been born into a well-to-do family and had lived the life of socialite until then, not needing to work to support herself. She worked very hard in the hopes of getting herself and her three daughters out of Germany, the country where the courts had permitted such unjust terms for the divorce. Five years of custody fights in the courts followed the divorce.

In 1955 my mother managed to obtain an immigration visa to Canada. She left by herself in August, when I was thirteen. All three of her daughters were left behind in Germany with her parents.

My father had always been fond of me and usually relented when I really wanted something. In October 1955 he gave me permission to visit my mother in Canada. I left as soon as possible in November for fear that he might change his mind. Travelling alone, I took the train from central Germany to the port of Bremerhaven (near Hamburg) and then boarded a ship to Montreal in Canada. It was an eleven day trip across the stormy Atlantic Ocean. I knew some written English from school, but could not really speak English, nor understand it when it was spoken. But I managed to find a taxi to take me and my trunks from the docks in Montreal to the train station. There I bought a train ticket to Toronto where my mother was waiting for me. She was surprised when she saw that I had brought several overseas trunks of her silver, china, oriental rugs, and books. She had not taken these with her when she had departed into the unknown. She was living in a single room, but now she could have at least some of her belongings there. My presence and the presence of some of her treasures helped raise her spirits while she went to secretarial school and sought a job. We used the trunks for tables and ate our simple meals using the fine china and sterling silver cutlery, sitting on an Oriental rug while illuminating the room with candles in sterling silver candlesticks.

In Germany my academic performance had been poor, but I was always full of jokes and was the social leader of my class. Having fun at school distracted me from the stress at home and my parents' struggles. I used to organize after-school events for our class such as going swimming in rivers or hiking in forests. My classmates had loved this. When I left for Canada, they in turn had arranged for a band to play for me at the train station as I departed. They had the band play an old-time song about leaving one's hometown but coming back again, "*Muß I denn zum Städtele hinaus.*"

Although I was only supposed to come to Canada for a visit, it had been my intention to stay with my mother — and I did. My goal at that time was to get out of school as soon as possible to help earn an income to support my younger sisters when they managed to come and live with us. My academic background was from a tough German school system and I was able to persuade the principal at the Bloor Collegiate High School to allow me to skip three grades. This was the first time in my life that I actually worked hard at school. The teachers were so kind and helpful. I couldn't believe it when all of a sudden I was considered a top student and began to win prizes. I was surprised to learn that there were monetary awards with these prizes. This money let me earn an income while still in school. I also had after-school jobs to help earn income, but had not expected to earn anything from my academic efforts.

For the first time in my life I began to enjoy academics for their own sake. Physics and mathematics were my favorite subjects. I focused on my studies, becoming much quieter and more studious — I was no longer the life of the class. I was shocked to see my picture hanging on the walls in school and in articles in Toronto newspapers. I continued to win many awards and scholarships, which paid for my tuition at the University of Toronto with money left over. I no longer needed the low-paying after-school jobs. I continued to win academic awards that helped support me and my family throughout my years at the University of Toronto.

I was in the Physics and Mathematics program. There was a theoretical physicist who came from the University of British Columbia (UBC) in Vancouver to give a seminar at the University of Toronto when I was an undergraduate there. He was incredibly dynamic, and inspired me a great deal — I even seriously considered going into theoretical physics. His name was George Volkoff, whom I learned later was Robert Christy's closest friend from their UBC and Berkeley days.

For my graduate work I had considered either going into theoretical physics at the University of Toronto, or else going to Princeton to study with Martin Schwarzschild, who specialized in stellar evolution. I decided to stay in Toronto because I was engaged to a graduate student there in experimental physics. Instead of theoretical physics, I was persuaded to study astronomy and astrophysics, obtaining a Masters in Astronomy in 1965 and a Ph.D. in Astrophysics in November of 1968.

Hearing the Christy Name for the First Time

I first heard the name of Robert Christy in the fall of 1963 when I was 21 and had just begun my graduate work in astronomy at the University of Toronto. Two of the professors specialized in variable stars, and Professor Christy had just made a major breakthrough in the understanding of the most important types of variable stars (namely, Cepheids and RR Lyrae stars, as described in Chapter 10). To me, however, Christy was just another prominent name in astrophysics at this time.

The summer of 1964 was the first time that Robert Christy and I were both at the same place at the same time. The occasion was the International Astronomical Union (IAU) conference in Hamburg, which lasted nearly two weeks. We both have vivid and pleasant memories of that conference but do not remember encountering each other there. This is not too surprising. Not only were there were several thousand people at that conference, but in addition Robert Christy was attending and speaking at sessions on theoretical astrophysics while I primarily attended sessions on observational astronomy.

Despite the fact that I did not encounter Robert Christy there, this conference did have a major impact on my later life. I met Professor Rudolph Kippenhahn there and was immediately interested in working with him. He was a theoretical astrophysicist who worked on stellar evolution and stellar interiors at the University of Göttingen in West Germany. I told him that when I finished my Ph.D. at the University of Toronto I would try to move back to West Germany to work with him as a postdoctoral fellow for a year — which I did in December of 1968. I also met Professor William Fowler at this conference and was likewise drawn to the idea of working with *him*. Later, in 1971, when I was hoping to come to Caltech, it was he who gave me my first position there. As it turned out, he was also the one who stood in for my father to give me away at my wedding to Robert Christy in 1973.

A Brief Encounter with Robert Christy

Robert was an invited speaker at the University of Toronto's June Institute of Astrophysics in 1968. At this time, I was a graduate student there, in the process of writing my Ph.D. thesis. I was still engaged to my first fiancé, but had come to the firm conclusion that I wanted to break our engagement. There were many people in the audience as Robert Christy gave his lectures describing his work on variable stars. He presented beautiful diagrams showing his calculated light curves of these stars, but I watched him rather than the diagrams. I was immediately attracted to him. He of course was not aware of me in the large audience.

Juliana Sackmann at her Ph.D. graduation in Toronto in November 1968

There were four days of lectures. Some of the young visiting astrophysicists from Harvard University asked me where one could go to have some fun in the evenings after the lectures. I mentioned a number of places that I knew of, including a basement where there was live music and dancing with psychedelic lighting. Catherine Cesarsky and her husband Diego were among the Harvard graduate students who decided to go to this music and dance place. Catherine approached me to say, "Our little committee has decided that you should ask Christy to come." I said, "No; I have a crush on him," but Catherine answered, "Well, I can't do it, I'm a married woman. You're the only one who can do it." Very reluctantly, I agreed. I was very surprised when he, in his quiet manner, agreed to come. When the group got to the basement with loud music and psychedelic lighting, Robert told me that he didn't know how to dance. I told him, "No problem, we'll go to one of the corners where we won't be seen and I'll show you how." And this is what we did. That was our only interaction in Toronto.

Engagement to Peter Biermann for a Year

I obtained my Ph.D. in Theoretical Astrophysics from the University of Toronto in November 1968. I received a tenure-track job offer from the Department of Astronomy at UBC, and one of the attractions for me would have been George Volkoff's presence in their Physics Department. At the same time I received a NATO Postdoctoral Fellowship that would support me generously for two years in any NATO country. I considered using these funds to go to Princeton or to West Germany. In the end I chose to go to West Germany.

One of the reasons why I chose to go to West Germany was to put some distance between myself and my ex-fiancé, who was planning to go to UBC and who would have visited me at Princeton. Other reasons for going to Germany included wanting to work with Professor Kippenhahn and getting to know my birth country. I also wished to use this opportunity to travel all over Europe while I was still single. I hoped to have children one day and would then no longer be free to travel.

When I arrived at the University of Göttingen in December of 1968, I met a brilliant young astrophysicist, Peter Biermann, who stood out amongst the others. He was not only very charming and accomplished but also was very open-minded. He spoke many languages and had travelled extensively. He had worked in Israeli kibbutzes for several summers, and in the astrophysical community in Paris. He was a graduate student of Professor Kippenhahn and was close to finishing his Ph.D. thesis.

A romance developed between us and in early 1969 we became engaged. Peter Biermann was willing to live in North America, and I was willing to live in Germany. I felt a connection with his mother and sisters, and with many of his acquaintances in the field of science. Peter's father, Ludwig Biermann, was a distinguished professor in astrophysics and co-director of the Max Planck Institut für Physik und Astrophysik in Munich with Werner Heisenberg.

In Nazi Germany, Heisenberg had been in a position similar to that of Oppenheimer in the U.S., namely, he was in charge of the most important of the six institutes that were supposed to develop nuclear power for the Nazi war effort under the auspices of the military (as is discussed in more detail in Chapter 5). At the end of the Second World War, as the country lay in ruins, the families of Biermann, Heisenberg, and Carl Friedrich von Weizsäcker had shared a triplex in Göttingen. There had been many discussions among those families of what had really gone on during the war. The children had heard some of these discussions, and it was there that Peter had learned many details of the German nuclear project. Later he shared these with me (as mentioned in Chapter 5), and I learned much that I would not otherwise have been aware of.

Peter and I travelled together to Canada and the U.S., and he also introduced me to his world in Europe. However after about a year, in early 1970, we decided to break our engagement, although we remained friends.

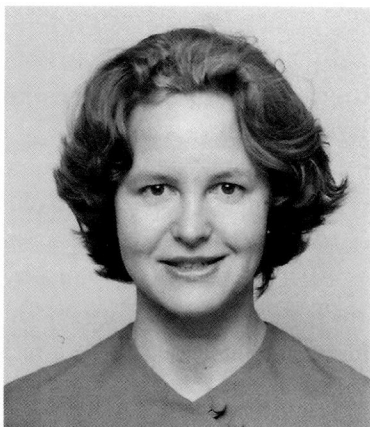

Juliana's passport photo

A Second Encounter with Robert Christy

In the summer of 1970 I travelled from West Germany to attend the eleven-day meeting of the International Astronomical Union (IAU) which took place in Brighton, England. Robert Christy was one of the speakers, presenting his work on variable stars, and I was in the audience. Again I felt a strong attraction to him.

One evening Ian Roxburgh, a theoretical astrophysicist who lived near Brighton, invited 60 of the 2000 attendees to a party at his home. Both Robert and I were in this group of 60. That is where we really first met. We spent the whole evening in the Roxburgh kitchen talking about rapidly rotating stars (my Ph.D. thesis), and about Robert's early life in Vancouver, Canada and his later life as a graduate student in theoretical physics with Oppenheimer in Berkeley, California. I did not sleep that night: I was sure that this man would be important in my life. I walked down to the ocean to watch the huge waves breaking on the rocks of the shore. I felt that

Robert's passport photo of May 1970

Robert would be like one of these waves, crashing into my life.

Shortly thereafter there was a train trip organized by the IAU where I ended up in the same part of the train as the delegation of theoretical astrophysicists from Portugal. There was a lively discussion, and they informed me that there was a ten-day astrophysics meeting in Portugal immediately following the IAU meeting. I knew that both Christy and Kippenhahn had chosen to attend that meeting, but I had planned to go to a ten-day astrophysics meeting in Greece (as I had done 1964) and had already paid for the travel, hotel, and meeting registration fees. The next day the Portuguese delegation found me and invited me to join them in Portugal for their meeting, offering to pay for the meeting registration and hotel expenses. I then searched for Robert Christy among the IAU attendees and finally found him, telling him of this invitation. He did not encourage me to go to Portugal; he didn't even blink an eye. I concluded that he was either married or not interested. I told the Portuguese delegation that I would not be coming since I had already paid so much to travel to and attend the meeting in Greece.

The next day the Portuguese told me that they would pay all expenses, including the air travel. I was still uncertain whether to go to Portugal. I was more interested in the topics of the meeting in Greece, and Christy had not shown any interest in whether or not I would be in Portugal. So I looked for my boss, Kippenhahn, and asked his advice as to which meeting to attend. He answered

"Portugal." And so I decided to follow his advice and go to Portugal — but ignore Christy.

A Romance Begins

After the Brighton meeting I went back to Munich to pack for the warm Portuguese climate, arriving in Portugal after the meeting had begun. When I arrived in Portugal there were great festivities going on at a castle lit up with torches and candles. Ambassadors from many countries were present to greet the astrophysicists coming from their respective countries. I could not help but notice Christy, whose height made him stand out, but ignored him as per my previous decision. The Portuguese meeting was going very well, and I kept company with my German and English colleagues.

At one point I was sitting by the pool with Kippenhahn and the German group, wearing a huge white hat to protect against the sun and a polka-dot bikini. To my total shock Christy approached me from behind and had the nerve to lift my hat and ask in public, with my boss (Kippenhahn) and all my German colleagues watching, "Who's hiding under this hat?" He asked if I would join him for a walk along the sand dunes by the ocean. I didn't know what to make of him. We walked for a long time and not much was said. The next day he asked me again to join him on a walk, and again not much was said. But eventually on one of these daily walks he did break the silence. Out of nowhere he volunteered the information that he had two sons, aged 26 and 24. I said nothing. On the next day's walk he stated that his wife and he had split up, that she had moved out, and that they were getting divorced. Again I said nothing, but I concluded that this meant that I could have a clear conscience. I would not be breaking up a marriage.

Later Robert asked me whether I had ever surfed, and I answered "no." I had not lived near an ocean for most of my life. He said he knew how to body-surf from living in Los Angeles and asked if I would like to learn. I enthusiastically answered "yes," and he taught me in that cold Atlantic ocean. The ocean was so cold that Mrs. Kippenhahn, who had gone in the day before, claimed it had caused her kidney infection to flare up. Robert and I decided to try it anyway. I was surprised what a strong, boyish swimmer Robert was, and yet that he always looked like a statesman even in the middle of the huge waves. He taught me how to ride the waves and how to dive under them when they were too high. When we emerged from the ocean, Robert suddenly said, "You look so beautiful like

this." Shortly afterwards he told me he was fifty-four years old. I asked how old he thought I was. He answered "twenty-nine," and I straightened up and said "close enough; I'm a full twenty-eight and a half." He quietly took my hands, saying nothing.

A group planned to see a

Juliana Sackmann and Robert Christy on the beach in Portugal in September 1970

bullfight and asked me if I had ever seen one. I said that I hadn't. Once I learned that in Portugal the bull did not get killed, I agreed to go. I invited Robert too, and we went to a near-by village to see a bullfight with stirring music and performances.

Later I asked Robert if I could teach him dancing again, as I had done in Toronto, and he said "yes." As we danced in the basement of the hotel it seemed to me that all of my colleagues had come to watch, pointing at us and smiling. (It turned out later, when I returned to the Max Planck Institute in Munich, that everyone seemed to know about our dancing, even visitors from England, Scotland, and elsewhere.) I was worried about all the curiosity of our colleagues, but Robert stated that it didn't bother him; he had led a very straight-laced life.

From that point on Robert and I tried to be together at every meal time to get to know each other better. Robert asked me what I thought love was like, saying that he had never experienced it. (I gave him a half-baked answer as I hadn't really thought about it.) At one point he said that he wanted me to keep myself free and go out with others because he was too old for me.

During the next dinner there were performances by local dance groups on stage. Robert asked me if I would join them on stage. I declined, but a little later the dancers came to the audience asking for volunteers to join them on stage. My colleagues from Germany urged me on, and I agreed. Robert was radiant with joy to see me dancing and our colleagues applauded enthusiastically. Later Robert told me that he loved to dance. He said that he had taken dancing classes as a graduate student at Berkeley, but had not been able to dance since then because his first wife did not like to dance.

Portuguese nuns came to the hotel to sell beautiful hand-made baby clothes. My sisters in Vancouver had young babies and I wished to buy some of these clothes for them, but was unsure about sizes. I asked Robert to help pick the right sizes since he had been a father already and had two sons. He came and was very helpful in the selections. Thereafter he bent down and kissed me, and asked if I would join him in his hotel room. I said, "No; it would not be good for you, it would not be good for me, and it would not be good for us." He was clearly upset but he said, "You're probably right."

On the last day, Robert gave up his lunch, to spend the time swimming with me, riding the waves at the beach. This was more of a sacrifice on his part than one might think. Ever since he had been hungry so often while a graduate student at Berkeley he had really loved and appreciated good food. Following this ocean swim Robert told me that he was in love. He said, "You are so sweet, graceful, and chic."

Robert told me that he and his first wife had known each other mostly via letters over the years before they married, and when they got together it was very different. He was not interested in carrying on a long-distance romance again. He wished me to give up my job and move immediately to Pasadena, and offered me his empty guest house. I said that I wouldn't come without a position in Los Angeles. I wanted to continue in my profession and to be self-supporting financially. He said that I would not be able to work at Caltech because of the nepotism rule, and that UCLA would be too far away. I said that I was preparing for a tenure-track position at the Observatory in Hamburg, and that if I couldn't get a position in Los Angeles then he should get a position in West Germany. I also told him that if he was Provost and Academic Vice-President of Caltech, and had the power to change the rules, then he *should* change them.

As we left Ofir, I asked Robert if we could stay in contact and Robert enthusiastically said "yes!" Robert again suggested that I move immediately to Pasadena and live in his guest house; he could help support me financially. I told him that this was something I could never do. We parted tearfully at the airport, Robert flying back to Los Angeles, and I to Munich. Just before we parted Robert said, "You continue to amaze me. You are one of the nicest things that have ever happened to me."

A Difficult Struggle — 6,000 Miles and 26 Years Apart

Three-foot-long roses arrived at my rented room in Munich soon after I had returned. For the next year many letters went back and forth between the two of us. Excerpts from some of the letters from Robert tell the tale of an irresistible attraction and the almost insurmountable problems — though in the end it seemed that fate was on our side. These letters illustrate Robert's personality and what was important to him. They also show his integrity and dependability.

> *Sunday Sept 13, 1970:* I can still feel your hands, see your face, and hear your voice. I hope I always can. It was a wonderful, wonderful thing to happen to your Robert
>
> *Sunday Sept 20, 1970:* It seems so strange — we were <u>together</u> for less than a week and now I want so much to be with you.
>
> *Wednesday Sept 23, 1970:* A day or two ago I saw a half-hour skiing film on TV and I was immediately transported to the mountains. I could almost feel the exhilaration of a good ski run. The feeling of freedom, the wind in my face, and beauty of the surroundings. I get a very special satisfaction in such circumstances. Somehow I can indulge in a certain wildness that I cannot express at home. I felt somewhat the same in the surf at Ofir [in Portugal] but not so much so.
>
> *Friday Sept 25, 1970:* I am overjoyed to hear you say that I am the right person — I hope it works out that way. Knowing you has made me feel something very special that I have not known for a long time — I do hope you are right.
>
> …I guess you refer to my thought of learning German. If you like the idea I will certainly have a go at it. I visited Berlitz school the other day to make inquiries

and explore. They mostly discussed *Bleistifte* [pencils] and *Tischen* [tables] etc. I am still looking for the most practical approach. I asked a colleague and he jokingly referred to Richard Burton who had learned a new language with each new mistress. This was not quite my intent of course.

I had mentioned to Robert that one of the reasons I wished to stay in Germany was that I loved my mother tongue and wanted to speak it as much as possible. I had not asked Robert to learn German, but it meant a tremendous amount to me when he volunteered to do so. Considering how busy he was in his profession and how difficult it would be for him to learn a new language at his age — not to mention the fact that he would have little opportunity to practice it — I found this irresistibly endearing.

> ***Monday evening Sept 28, 1970:*** I have registered for a German course and start tomorrow at 8 AM. I will see how it goes and whether I can make progress this way. Do not expect miracles. This is no doubt the first time the Provost here has ever taken a course. (I am really not <u>registered</u> but merely have arranged to take part in it.)
>
> …I want you to see me where I am and get to know the people here.

I was working long hours seven days a week at the Max Planck Institute to finish several research projects before December 1970. My position there would terminate then, and I would have to find a position elsewhere — either with Prof. Kippenhahn at the university in Göttingen or with Prof. Weigert at the Observatory of Hamburg. I was therefore trying to cut down on my letter writing, which was taking up so much of my time and energy.

> ***Friday evening Oct 2, 1970:*** I am trying to do what you say, namely write once a week. I will do my best but last night and the night before I spent all evening just thinking about you. You kept popping into my mind from all directions and there you remained. I do not know any way to solve this problem except to be <u>with</u> you.
>
> …What makes me worry is the problem of a research position here at Caltech. I told you that we have <u>no</u> theory group on stellar interiors. Prof. W. A. Fowler (a nuclear physicist and a very old friend of mine) usually has a group of theorists which includes one or two interior people — Weigert was here, so — I believe — was Dr. Thomas. You might casually ask him what he knows about Caltech — I am so relieved that after telling me how much you enjoyed your discussions with him you were so kind as to tell me that he is married. I need not say more to tell you my problem and I am happy you are sensitive to it.
>
> …If nothing else can work I could seek research money myself and employ you but I am not sure my conscience would permit it.

I am frankly very worried. I recognize your need to have a position and I fear it may prevent us being together. Please tell me if you think of any possibilities I overlook. We will see each other in a few months and in the summer. Beyond that we must do everything possible and also trust that fate will be on our side.

...I will also see if I can find any peculiar, little known, grants you might apply for. Trying to get you here may make me devious and that is <u>not</u> my usual approach.

I agree so much that what we have least of is years. I do not want to wait till next fall to have you here. Please write very clearly and frankly of your views on this question as I agree what we do now is most important.

...You ask what is wisest for us to do now. I do not know everything that is on your mind but to me it is important to see as much of each other as possible in order to be sure the feelings I feel — and I hope you also feel — are real and not conjured up out of an imaginary world of Ofir and letters. If we find when you visit in ≈ Feb ≈ that we still feel the same (and I do so hope it is so!). Then we must make every effort to be together in the summer and find <u>some</u> way for you to have a professional life if you were married to me. I would like to speed this process up and see you also all spring but, unless you just decide to come here anyway, I don't see how to arrange it. Please write me very frankly about your views on all of this. I find your basic instincts about us very sound and am quite prepared to see other sides of the problem if you will point them out.

I will now wrap up all this and only those pictures that fit in an envelope. The others (including those of my earliest years) will follow. One picture I am also sending is of me participating last spring in a

This is the "standard ski race" photo from Aspen, Colorado that Robert refers to in this letter of Oct 2

"standard" ski race at Aspen. Experts run the course and determine a standard time. Then anyone takes a run and if your time comes within various standard percentages exceeding the expert you may get a gold, silver, or bronze pin. I am inordinately proud of getting a bronze pin in that race — that is why the picture is important to me. It does not show me in any readily recognizable form but it pleased me very much. Do not lose my <u>only</u> pictorial proof that I can ski.

Sunday Evening Oct 4, 1970: I find I must talk to you in the only way I have so I will write a little when I feel in the mood but will wait perhaps till the end of the week to send it. In this way it will be easier for you.

...You may (since you have good instincts) or may not realize that I am basically quite shy. I was very much so when a young man and much of it has worn off but still some remains.

...I sometimes wonder in a perverse way if our relationship is all planned to make me eat my own principles about the employment of spouses. It may come to that, that the only way I can marry you is to hire you. It sounds odd but I presume you understand my quandary.

Monday afternoon Oct 5, 1970: I am trying to slow down my correspondence, as you see, in order to make your visit here come sooner, but I think of you nearly all the time.

...I would love to show you the things I love and do with you what I love to do — and also to learn to enjoy your special pleasures and enjoy what is important to <u>you</u>. For years now I have had to live the important part of my life alone and I want so much to share it.

...My main problem is that I now appreciate also your needs to fulfill yourself. I am now much more understanding of people than when I first married. In fact that is perhaps the biggest change that has taken place in me in the last fifteen years. It is because of that, that I now can be Provost. Before, I understood too little of people — myself included. I will do better next time. I hope you can plan your future so that I have the possibility.

Wed. evening Oct 7, 1970: What you say in your letter about us [that Juliana wished to live as close as possible to where Robert lived] makes me feel as though I was walking on air. In the modern vernacular "higher than a kite". Do not worry about my patience or impatience. I can wait a few months if it is necessary but not indefinitely or if not necessary.

Friday evening Oct 9, 1970: In only a few hours, it will be four weeks since we parted. In many ways these weeks have been the most significant in my life. It may be three more months until we meet again — it will be a long time!

Saturday afternoon Oct 10, 1970: I am so happy that you like chic clothes and I want you to have them. You dress so attractively — it is one of the many things about you I like so much.

...If I can stay awake late enough I may telephone you tonight. But I have been awakening at 5 AM so I may not make it. I want to hear your voice again and tell you what you must already know — I love you — and emphasize how important it is for us to be together.

Wednesday evening Oct 14, 1970: Only rarely in my student days did I feel relaxed. Youth is a difficult period. I relax much more and more readily now than then. You mention that from some points of view it would be best if you could come in the spring — right after the MPI [i.e., after Juliana's position at the Max Planck Institute ended]. BY ALL MEANS!! I feel a quiet desperateness thinking about how long it may be. … I am very well aware of the years and I am very anxious for us to get an early start on what I hope will be a LONG life together. If taking your hand now would bring you here, I will fly over to Munich and bring you back. I have felt, however, that <u>you</u> needed an assured position before coming. If I have misunderstood you, you must correct me. I will give more thought to this and see if something can't be done (short of abducting you which I am tempted to try!)

En route to Chicago, Thursday afternoon Oct 15, 1970: I appreciate that anything less than a regular appointment could be unsatisfactory to you except perhaps that you might come a month or two before the start of a regular appointment. I am simply supposing that you think very much as I do on questions of this kind. It is one of the many things that attracts me to you — our basic values and instincts seem to be rather similar. So far I have found that I was in basic agreement with you — even when you may have thought otherwise. This doesn't always make our course easier, however, since we are both, perhaps, rather stern taskmasters.

… In case you haven't found out, after you — I love the mountains. The Sierras, the Rockies, the Alps — I have visited them all and climbed in all of them. I plan to do more.

Photos of himself that Robert included with his Oct. 17 letter. Note that the photo above shows Robert with a new tie, rather than the New Mexico tie that had characterized him for decades (see the "Necktie Story" in chapter 9).

If & when I get to Munich I assure you I want to spend some weekends in the mountains.

…I agree with you regarding past photos as for now, though the emphasis of my reasons is somewhat different. In my case, I have had more than enough time to adjust to my own past. It has taken many painful years to reach the place I am now at! However I do not want to inflict that on you. … As for your family — there the timing is up to you. Clearly it would be easier after we understand better our relation to each other. But sometimes external circumstances dictate these things and I am happy to adjust to these. Perhaps, in fact, my occasional emphasis on "fate" is merely a willingness to adjust to the apparent boundary conditions of the external world while at the same time I do my best to influence the world in the direction I see.

En route Chicago to Los Angeles, Friday Oct 16, 1970: For several years now I have taken no pictures because I did not really want to be reminded. When you come it will mark a new departure and life will start again for me! I will want to get a new camera to celebrate the occasion and to record us.

We have just been crossing the West which is MY country. I love it all. … I know all this region from the air and

More photos that Robert included in his Oct 17 letter. Above: Robert (center) meets Rudolph Kippenhahn (right) and Alfred Weigert (left) while attending a meeting on variable stars in 1966 in Bamberg, Germany. Below: Robert meets an onion lady while in Bamberg.

some of it from the ground. I want to show it to <u>you</u>. I hope you will also be able to show me the places and things you love! I want to partake of your life just as you partake of mine. That way we both will be enriched!

Saturday afternoon Oct 17, 1970: I am enclosing some pictures. Yours I return with regret — I do like them but what I most want is you! I also enclose two from the 1966 Bamberg colloquium [in south Germany]. The others are samples from those I took a week ago here for you. I am only partly satisfied with my efforts.

Saturday afternoon Oct 24, 1970: I feel that I should be with you, to hold you and help you in the very difficult problems you are facing. Almost as much from what you do not say as from what you say, I sense the manifold conflicts you are facing: To visit here at Christmas or not, to take a position at Hamburg or not, to go to work at Göttingen with K [Prof. Kippenhahn] or not. And related to all of this are uncertainties about Pasadena and, perhaps, me. I do not have answers to your dilemmas but I do feel that we should try to be as close as we can in trying to find the answers. As you well know, I am not an uninterested bystander in all this, I have a tremendous stake in your decisions and I want to participate.

…Brown goes away next Sunday for three weeks in Helsinki and Strategic Arms Limitation Talks. During most of November I will be acting President. So you see why I am only able now to answer your letters.

Before I met Robert I had been considering taking a long-term position either with Kippenhahn at the University of Göttingen or with Weigert at the observatory in Hamburg. After I met Robert I came up with the idea of applying for the Senior von Humboldt Award, which would allow me a short-term position anywhere in Germany. This would let me stay in Munich (saving the trouble of a major move) and would also leave me freer to leave if I could obtain a position near to Robert.

…You mention your discussions with Kippenhahn and that he wants you to work with him but you do not say whether you may or not. I suppose it depends on von Humboldt. (It is clearer now that you have told me what you are up to — I get very confused when you keep too many secrets.) You say one thing which I must react to — you mentioned to K the possibility of coming here in fall 1972. I can only say that as I view the difficulty of surviving a few months away from you I think that fall 1972 is at <u>least</u> one year too long for me. I cannot think of any circumstances where that would be reasonable. If after next summer we still feel as we do, but presumably much more so, then it would be unreasonable (for me at least — perhaps you view things differently) to wait a year <u>away</u> from you! If we do not feel that way then the <u>purpose</u> of coming here (at least my selfish purpose) is gone. So much for 1972 — I will have none of it!

…I have begun to enquire about a trip to Munich. Roughly I think of about Dec 18 to Jan 2. I do not yet know if this is possible for me but what I can make possible depends also on your reaction — would you like me to try? I begin to realize that next July is too far away for me and if you can't come here in between then I must go there. If you were able to accept a position here in Feb or March it would be different but you say nothing about whether you will try or not.

Sunday morning Oct 25, 1970: I was interested in your comment about weddings at Ulm — do you think perhaps you would like to be married there? I had thought you would prefer where your family could be present. If it were to be me, I would be happy anywhere that provided that it is with <u>you</u>.

…I have thought more about the question of starting a new family and have found reasons why it could be a good idea from my point of view [it turned out later that these reasons were not enough to really convince him]. I am, of course, fully aware of your hopes in this regard.

Monday evening Oct 26, 1970: I would love to see John Cranko's ballet with you! I have seen but little ballet but what I have lacked is someone to go with. Many things like that have been missing with me because I would have had to do them alone. I also want to do the other things you like — such as shopping in Munich. Two can enjoy that even more than one.

Sunday evening Nov 1, 1970: I notice you look up some words in the dictionary to get the correct English: I can also translate from the German so you may, if you wish, write me in a mixture with some German. Please, however, write more clearly if in German since I do not automatically recognize unfamiliar words as I can in English.

You ask how your yellow rose petals arrived. — Beautiful, even now there is a faint fragrance and there is only partial discoloring even now. Which reminds me how happy I am that you like beautiful things and flowers. I need someone with such taste because I also like these things but never seem to seek them out for myself.

…I have hinted only to Harold Brown that I might want to go away to "see more of someone I met last summer". He understood and wished me luck. Well so do I (wish me and us luck).

Monday evening Nov 2, 1970: I have also noted what you remark on in your note of last Tuesday (Oct 24). In many cases we seem to touch on the same topic in letters written at the same time which cross. I believe another example or two will have reached you already. It certainly shows that our minds are travelling along similar paths!

Wednesday morning Nov 4, 1970: I explained before that although I sometimes work long, it is not to me stressful if I am not fighting a deadline and I also relax with music, or playing tennis or something. I sometimes play some old records I have or sing some songs or listen on the radio to music.

…On telephoning to you, I have still not received a telephone bill including charges for our first talk. Perhaps they couldn't believe it.

Saturday noon Nov 7, 1970: I now know how much my telephone bill is. One just came that included three of my calls to you! But it is worth it.

Sunday evening Nov 8, 1970: It was wonderful last night to talk to you, to hear your dancing voice feeling happy — you sounded wonderfully happy to me — and to hear that you also think we might meet at Christmas. It was remarkable to me that I went to bed and then felt within me such a strong urge to talk to you that I could not stay in bed so I got up and called. Even though I cannot believe in ESP I think it wonderful if we are so well tuned that we are feeling similar things at similar times.

…Last night I played some records and sang some songs. When I come to Munich I would like to find some records of German songs (not modern popular, but traditional — "folk" songs). I would also like to find a song book so I can learn such songs for myself. I have only a very few German songs in my books.

Monday evening Nov 9, 1970: Your letters of Thursday & Friday came today and begin to suggest your rethinking about Christmas. I am very happy you see it more as I do.

Tuesday evening Nov. 10, 1970: I am very glad that you are willing like I am to be a little flexible on our exact plans so that we can be more adventurous. This is the way I usually travel since it removes the iron bound schedule. Sometimes, however, one is less comfortable.

Thursday evening Nov 12, 1970: Yesterday I was so frustrated! It was a special National holiday — related to "Armistice Day" so no mail. I knew your letter was there and drove to the mail station but it was locked and empty so I could do nothing. Today it came and with such a wonderful warmth. I would love to help you celebrate Christmas in your traditional way! It sounds wonderful!

…Your long letter of last Sunday makes me feel even more, if that is possible, that I want <u>you</u>. I love your picture of the traditional Christmas season. Is your mother responsible for your warm memories of *Heilige Abend* [Christmas Eve, which is the traditional Christmas celebration in Germany]?

Saturday 1 PM Nov 14, 1970: I have just been watching the world championship gymnastic competition on TV (taped) from Lubleana Yugoslavia. It is fantastic. The men are tremendous and the women beautiful! The grace of their movements reminds me of you in your rhythmic exercise movements.

Monday evening Nov 16, 1970: I have just realized that I have not told you enough about myself so that you would know if I preferred a big swinging resort or a quiet country spot. I will try to get it straight when I call you again tonight to settle our *Weihnachts* [Christmas] plans. When I go skiing I normally devote nearly all my energy to skiing and do little or nothing in the evening. In part, of course, this is because I am frequently alone. With you I would expect to do more in the evening but still I doubt I would very often stay up very late. My preference is no doubt to be in a quiet village with a country atmosphere rather than a major resort center. However in skiing one does not always have so much choice since the best places are also frequented by many people. But the night life is not an important part of skiing to me.

I am agreeably surprised that you also like to travel into the unknown. My usual procedure is that since I don't know all about where I go, I am prepared to adjust my plans as I go.

...I concur with your proposal to speak to me in German — I will have to ask for explanations of words but only in that way will I be exposed to enough German to really learn.

Tuesday evening Nov 17, 1970: I hope you understand my conscience in these things — I have a strong sense of responsibility which in general is not easily quieted and you should not suppose that I can always as easily leave my job. The need to see you, however, is rapidly destroying my usual reactions! Right now I feel wonderful about the thought that in just over four weeks I will be on my way! The plan <u>could</u> still be upset by emergencies but I am counting on it now.

Now that it is becoming real that we will soon see each other, I can confess that there are aspects of letter writing that worry me. I had never communicated thus before with any one and I now begin to find it almost an addiction! I worry that <u>letters</u> could become the primary object of interest, rather than the person.

Wednesday evening Nov 18, 1970: Your special delivery letter of Monday morning arrived at noon today. I am delighted that the flowers arrived. Sending them makes me so happy because of the fact that they please <u>you</u> so much. Giving is a pleasure when the receiver is made happy. I will accept the kiss when I arrive.

I am impressed by the effort you have put into getting us reservations but I'm sure the result will be well worth the while.

...It is so exciting to get your letters and to now realize we will soon be together. In a way I miss you but in many other ways I have only recently got to know you better so it isn't as much that I miss you but that I <u>want</u> you — really it is both missing and wanting.

Love from your Robert.

Saturday night late Nov 21, 1970: I have just returned from the party where I called you on the phone.

I called you because I was worried by the tone of your letter (or card) of last Wednesday. After talking to you I am still worried. You seem depressed to me and this makes me unhappy. I asked if I could do anything — no answer. I am sure that you realize that I feel close to you and if you do not feel right then I notice it too — just as you can sense if I am having problems. But what I feel most is the need to communicate with you and sometimes I feel that you do not want to communicate.

...Sunday morning

...Last night you said you wondered if we should write so much. I would not want writing to be a burden to you and I would be very happy to get brief notes or cards. I find that for me, writing gives me sometimes the illusion that I am with you and then I enjoy it. At other times I lack the illusion but want very much to communicate and I write, it takes me no more time away from other things than I would spend anyway thinking about you.

Monday evening Nov 23, 1970: I will send this by Special Delivery as an experiment to see if it saves time. I have no essential news to report however — except that I want you for Christmas and for always.

…I would love to see *Das Münster zu Ulm mit dir* [the Cathedral of Ulm with you]

Dein [your] Robert

Mittwochsabend [Wednesday evening] Nov 25, 1970: Before I come, you must tell me if you prefer the German or English pronunciation of Juliana. I am assuming the German is Yuliana — I have not asked.

Donnerstagabend [Thursday evening] Nov 26, 1970: I want to respond to your very warm letter of last Saturday. You say that as you were shopping you felt such a strong urge to write to me. As you now know, I, at the same time was trying to telephone you but found you had gone shopping! If that is the way it was, I do not explain but am very happy that we communicate so well.

Saturday noon Nov 28, 1970: I have in recent years more or less consciously adopted a new view. When I was younger I also pushed very hard, since I was aiming at the future. I have now come to realize that I must live for the present and not for the future — the outlook changes with age. Consequently I try hard to mix into my life things that I enjoy even though it means putting off some other things — such as research now — that I might do. I can no longer devote myself 100% to science so I am willing to do some when I can but not to displace various forms of recreation that I think I need.

Sunday noon Nov 29, 1970: I have devoted some thought to a question of timing (as perhaps you also have). It is undoubtedly premature and presumptuous but, nevertheless, under the circumstances it is relevant. If our meeting this Christmas is successful (as I hope) and if you are able to pay a visit here which also leaves us planning for the future, then I think it would be most important that we be together as much as possible. This should start next summer. But it should <u>continue</u> in the fall and throughout the year. During the period we are together, starting sometime in July (if all goes well) we should both be able to come to long term decisions and, at sometime in this period, we could be married. There — I said it—. I do not want now to try to be more specific — it would not be too meaningful — except to indicate where my own hopes lie.

…Goodbye for now Juliana *mein Schatz* [my darling]

Dein [your] Robert.

Wednesday night Dec 2, 1970: I delight in your interest in shopping. It means so much to me that you like to do these things. Your approach is very much like mine — buy only what I would want — but in recent years I have not felt like doing it. I used to do all the Christmas shopping for the boys. Now I want to do it with you when I can find some time. — I love the little ribbons you use sometimes in your letters. There are so many touches that mean so much.

...How to arrange for you for next year I do not know — I only pray.

...I realize, as no doubt you also do, that our meeting is important for us. At times I am almost afraid — but I am so anxious that it should be successful and so it will be. I long for you now and always my sweet.

Thursday night Dec 3, 1970:   Last night ... you said something which suggests what I often suspect — in many ways you have more sense than I — at least about me. Perhaps about you, you should take some of my advice.

...I am inclined to purchase something of intermediate price and when I find out what I really want — then I know enough to spend more and get what I want. This is what I plan to do about the camera. If I get the bug then, with your help, we can investigate the kind we really want. (Do you think that, having been married once I am now in a better position to find a first quality wife for the second time?) Remember I have always said my jokes are not always recognized. But they are not always said totally in jest either. In that respect I think I have found it but still haven't completed the negotiation (for the wife of course).

...It won't be long now but until then I do accept the kiss you offered a few days ago and I offer several in return from your Robert.

Mittwochabend [Wednesday evening] Dec 9, 1970:   I'd like to help you look for little boys' toys — I feel close to them of course for two reasons — I was one once and I have also seen them grow up. But I would feel strange looking for girls' toys as you may have initially looking for gifts for boys.

...Sometimes I am almost afraid of this visit because so much hinges on it. But that is the way it is and it is one of the problems of our relation which has been so much on paper and so little in person.

Samstagnachtmittag [Saturday afternoon] Dec 12, 1970:   I find very appealing, in an unusual and intimate way, your page written in bed *Mittwochmorgen* [Wednesday morning]. I like to think of you writing first on your tummy and then your back but I have not been able to distinguish the positions from the appearance of your handwriting. Nevertheless there is an attractive kittenish quality to your discussion there that I had not seen before. I have so much yet to learn about you at the intimate level and time is so short! (Also vice-versa.)

Robert's Visit to Me in Europe: Christmas 1970

I had been looking for any possibility of getting a position at Caltech. I had met Willy Fowler (a Caltech professor) at the IAU meetings in Hamburg in 1964 and in England in 1970. I figured that if I just sent him an application for a post-doctoral position it would just end up in the wastebasket, since he received so many such applications. Some of my Munich colleagues had been post-docs with him and had told me about him. So I made an expensive trans-Atlantic

Robert and Juliana skiing at Davos in Switzerland, around Christmas 1970

phone call to Willy Fowler, telling him that I would like to work with him as a post-doc and that I would fly to Los Angeles in the spring of 1971 so that he could meet with me in person. He was surprised that I would go to such lengths. I told Robert that I did not wish to be seen with him on this visit. I did not want to come to Caltech on his coat-tails.

Robert came to Munich to see me that December. We intended to travel to ski resorts and visit some of my friends. Robert and I went skiing in Davos, Switzerland, and later at a smaller resort in Austria. Robert was an excellent and fast skier while I was a slower intermediate skier, but we both still enjoyed it.

I again insisted on separate rooms. At the Munich airport as Robert was leaving for California he told me that it was over. This was quite a shock to me because we had had such a wonderful time together — every minute of it. I had been engaged twice before, but this relationship was more exciting and

Juliana at the airport, before she and Robert parted

fulfilling than either. This visit had confirmed for me that I wanted to see more of him. I had believed that the romance was just beginning and was seriously considering spending the rest of my life with him. When I asked Robert why he considered that it was over, his only answer was that he didn't like my style. It was only later that I realized that he was worried about my insistence on separate rooms, due to problems in his previous marriage (that he had not told me about at the time).

In his first letter after we parted, Robert wrote:

> As I think back to our wonderful two weeks together, I am still very much torn. My heart is very strongly committed to you but my head tells me to be cautious for the reasons — essentially threefold — that we discussed. Sometimes my heart is stronger and sometimes my head. In the time being I guess my feelings will stay that way. I missed you last night and I missed you even more this morning. … Juliana, I hope I can get myself straightened out so I can sign myself as <u>all</u> yours. … I am enclosing a Davos Derby brochure. It serves to remind us both of our stay there. I would like to return but in only <u>one</u> room next time. It was a fine place and excellent skiing and you were wonderful.

The threefold reasons that Robert mentioned in this letter had been discussed in the first few days of our meeting in Portugal the previous summer. Robert was concerned about our age difference of 26 years, fearing that I might later choose to leave him. He felt that I would want to have children while he had no such wish. Finally the fact that I had refused to share a room with him reminded him of the lack of closeness that had persisted in his first marriage (though he did not describe this last reason to me until much later). Three months earlier, in one of his first letters (October 2, 1970), he had written about the age difference: "As you must know I was very hesitant about permitting myself to become romantically interested in you because of our age difference and that you would not be interested in me. I still worry that you may vanish, abandon me, or what have you."

Juliana Visits Pasadena, and Takes a Two-Year Position There

Fowler was of course still expecting me to show up at Caltech, and Kippenhahn had advised me to try for the position and bring what I learned there back to Germany, so I came to Caltech in mid-January of 1971 to introduce myself to Professor Fowler. After an interview and a discussion what I would be able to do for him, he got up and said "you're hired." I later learned that Fowler had checked with Robert (who was the only theoretical astrophysicist at Caltech at that time working in a field similar to mine) whether he knew anything about me and my work. Robert had answered that he did not wish to discuss it — because neither he nor I wished him to use his influence for my application.

Incidentally I did not stay in Robert's guest house but at the Huntington Hotel (today called the Langham Huntington Hotel), a gorgeous luxury hotel embedded in the greenery of Pasadena. Robert offered to pay for it, saying that he had gotten extra income from some consulting that he had done. I asked Robert whether he could bring his sons home for me to meet them, as I considered that the romance was not over but would not wish to continue if his sons were hostile. I found that Robert's sons were adults close to my own age and that they were gracious gentlemen, not hostile at all. During this first visit to Caltech Robert took me hiking in many of the local mountains, into the National Parks in the desert, and to the most beautiful beaches around Los Angeles where he taught me more about body-surfing in higher waves and a warmer ocean than we had encountered in Portugal.

Again Robert and I had a wonderful time together during this visit. Yet Robert still considered that any romance was over, despite our continuing attraction to each other. After I had left Pasadena and returned to Munich, Robert's letters continued to speak of his wish to be with me. In a letter of

Juliana and Robert at California beaches in early 1971

February 10, 1971, Robert wrote:

> I still miss you very much but at the same time have my fears of going farther. Looking at the pictures [photos of Juliana] reminded me again of the happy times we have had together. I am very glad that you were able to visit here for two weeks. Every time we get

Juliana in California, on a rest break during a hike with Robert

> together it helps us to appreciate what being together means to both of us. … I hope you like the pictures of you as much as I do. I can still feel you close to me sometimes…

On the 16th of February, he wrote:

> Your proposed surprise has me all excited. When you first mentioned a surprise on the phone, it occurred to me that you might be planning a visit [for May] but I did not want to guess that aloud in case I was wrong. I am delighted at the prospect!
>
> Last night I was wakeful in the middle of the night for a few hours and you will not need many guesses to know that it was you who was in my thoughts. In my thoughts you were also in my arms — I can't get over wanting to hold you close.

and on the 20th he wrote:

> Last night I went to bed early — about 9:30 — and woke up early — about 3 AM. Perhaps I am trying to keep in tune with you as you go East. But it won't work any farther than Montreal. When you get to Munich we will have to sleep and

> wake on our own. But I can still try to dream and I can still imagine you are with me. I would like to find out if I can learn to sleep with you in my arms—. It is a delightful thought.
>
> Now I will say good night so that I may post this and it can start its way to the East. Your letter of Thursday from Vancouver lifted my spirits again and keeps you close to my thoughts.

Juliana and Robert on a picnic in 1971

In May of 1971 I flew to Pasadena to celebrate Robert's birthday with him. This surprised him a great deal because he was not accustomed to celebrating birthdays. To save money, I stayed in his guest house this time. Again we went hiking in a number of places in California, and again it could not have been a more

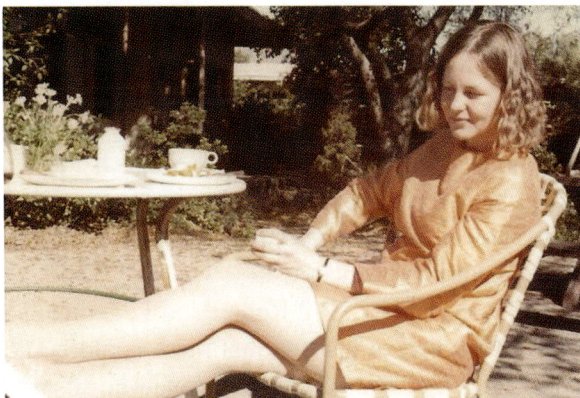

Juliana in Robert's garden in 1971

exciting and enjoyable visit. But again Robert stated at the end that any romance was over.

This time I accepted Robert's decision and decided that I would come to Caltech only for professional reasons, even though the initial impetus to apply to Caltech had been for romantic reasons. From that point on not many letters were exchanged between Robert and myself. In fact I was seriously considering not taking the Caltech job, but Professor Kippenhahn and other friends in Germany insisted that I not let go of the opportunity to work with Professor Fowler at Caltech. So I reluctantly arrived on October 28, 1971. Robert insisted on meeting me at the airport, but both of us were now in agreement that the romance was over.

Robert's Issues: Together But Not Together

I lived in the Caltech Athenaeum while looking for an apartment to rent. After the tiny spaces that I had lived in for three years while in Germany I was hungry for space, and was glad to discover that I could afford to rent a huge four-bedroom, two-story house close to the beautiful suburb of San Marino. However this house was totally unfurnished, and Robert, being the gentleman that he always was, graciously lent me some of the furniture from his unoccupied guest house. (Some time later, I gave up this big house, as it was too difficult to furnish, and rented a large and charming two-bedroom one-bathroom apartment in a duplex.) Also I hadn't driven a car regularly during the three years I had spent in Europe, and Robert loaned me his car so I could get some practice and regain my confidence until I bought my own car. In addition he took me body-surfing again.

We were still acting as if the romance was continuing, but Robert stated that it was over and that he was simply being a gentleman and a friend. He talked about seeing women his own age rather than someone as young as I. He certainly took a liking to my mother when she visited.

I made new friends while at Caltech and enjoyed their company as well. Some people visited me in the hopes of starting a romance. However when these men showed up at parties that I hosted in my apartment, Robert tended to get quite jealous and concerned.

Until the end of 1972 Robert and I continued to have fantastic times together, but Robert always insisted that there could be no future for us. After this long period together Robert finally opened up to me, this time in much more detail than what he had said when we were in Portugal. He said that he was worried about the 26-year age difference, which was not apparent at that time but which would show up later as he got old. He was worried that I might have affairs with younger men. He was concerned about the fact that I had not had any children, and would want children with him if we married. He had already had two children and did not want more. He recalled being up much of the night taking care of his sons when they were babies, and then heading in — sleep-deprived — to a long day of work. He never wanted to repeat this lifestyle. He wanted to be free to play tennis, go skiing, travel, and not be tied down by little children. He also stated that he was supporting his first wife generously and really did not have enough income to support a new family.

Figuring Out What Was Best for Him

My position with Willy Fowler would end in late 1973, and with it my work visa. So in January of 1973 I began to apply for other positions in North America, mostly on the west coast such as Stanford, the University of Washington in Seattle, UC Santa Cruz, and Simon Fraser in Vancouver. I told Robert that if I got one of these positions I would never return to Caltech, because I felt that we had gone back and forth long enough. Even though we had had a fantastic time together,

Robert on a local trip in 1973

he had stated often enough that he did not wish to get married. He had stated that he needed to marry someone his own age who would not want children.

I gave Robert two months, until the end of February, to figure out what was the best for him, before I would accept a position elsewhere. I tried very hard not to have any contact with him during that period. I told him he should see a professional counselor to help him resolve the emotional baggage that his first marriage had left him with, so that he could avoid projecting his fears (resulting from his first marriage) onto me. He agreed to see such a counselor.

Late in January 1973, he insisted that I accompany him, his best friend Jack Roberts (who wanted to meet me), and Jack's wife Edith, on a joint weekend skiing trip. During the five-hour road trip to the Mammoth ski resort, I was asked to sit in the back with Jack, who interrogated me for the entire trip about my background and my goals.

Later that weekend Robert came to me and told me that he had firmly made up his mind now. But I told him that I didn't wish to hear it anymore, and that he should take more time with the counselor. Though I did not say so explicitly to Robert, I feared that this would be like the previous episodes, when he would say he was certain only to have the ghosts of his past reappear.

At the same time, to heal my own wounds, I accepted invitations from others to escort me to various events, including ski trips — although I always would have preferred to be with Robert.

In the middle of February, Robert called again and said that he didn't need more time. He asked me to come to his house so he could tell me of his decision. I put on a dark dress as if in mourning, and prepared emotionally to never see him again. I arrived at his house, but once inside I stood close to the front door so that I could leave quickly. Robert told me to wait, that he wanted to get something from his study. He came back with a small white office envelope and gave it to me. In it was a gorgeous diamond ring. He told me that it had been his mother's engagement ring. He said in his quiet way that the counselor had urged him to "grab Juliana quickly, before it was too late." It was such a shock because I had prepared myself emotionally for the opposite. I simply answered "yes."

A Verbal Marriage Contract

At this point Robert wished to discuss what the two of us could do to have a successful marriage. This discussion turned into a verbal marriage contract. The terms may seem harsh, but it was the first time that we had communicated so explicitly about our respective worries and concerns.   This verbal marriage contract became the basis for a very successful marriage.

Robert said that he wanted a wife who would travel with him, especially on vacations, but also to meetings when possible; a wife who would join him in his outdoor activities and who would enjoy dancing with him; a wife who would enjoy entertaining and having fresh flowers in the house; a wife who would look after the cooking and do all the shopping.   He in turn was now willing to make a concession and agreed to have one child, provided that he had no night duties.   In addition he did not wish the presence of a child to prevent him from travelling with his wife, or to prevent him from enjoying outdoor activities — hiking, skiing, swimming, tennis, and the like.

In turn I told Robert that I wanted the possibility of a second child.   I proposed that if the first child caused no trouble, with no night duties or lifestyle limitations for him, a second child should be possible under the same conditions — I agreed to have no more than two children. I requested that there be no television in any room of our home except possibly his study. I made it clear to him that it was important to me that I continue in my profession as an astrophysicist.   It was also important to me that there be a "separation of estates" to separate his first and second family financially. I suggested that everything that he had produced and saved financially up to that time should go to his first family, and that the two of us would start from scratch to develop financial security for ourselves and our future children.   This would be difficult to accomplish because Robert was nearly 57 years old

Engagement photo taken in February 1973

with few years of earning potential left. I told him that I would put considerable effort into investments to help pay for our new family's expenses.

Robert had always been a most generous gentleman. He had already given all of his savings (except for his debt-free house) to his first wife at the time of their divorce. I suggested that his sons should receive their inheritance from him early, before our marriage: that he should give them most of the proceeds from selling his only remaining asset, his debt-free house.

Concerning the 26-year age difference and Robert's fear that I might leave him as he got older, I promised him that I would never do so. As far as the travel, outdoor activities, and lifestyle were concerned, I told him that I had the same tastes and wishes as he did and there was absolutely no problem with that.

This detailed verbal marriage contract, along with the solutions that the two of us had worked out in the three difficult years leading up to it, provided the basis for our long and happy marriage.

Robert sitting smiling (at right) with Reverend Brandoch L. Lovely and Robert's best man Jack Roberts (standing), at the Unitarian church in Pasadena just prior to the wedding

Not explicitly verbalized was my intention to make up for the unfortunate circumstances of his earlier life, to try to extend his life with me beyond the usual life expectancy, and to make him the happiest man in the world.

A Happy Marriage

Now that the uncertainties of the relationship were over, a new ecstatic change took place in both our lives. The two of us celebrated our engagement and announced it to relatives and friends. From that time on we allowed ourselves to be seen together in public in Pasadena and at Caltech.

In March we were noticed together in a café by the daughter of Harold Brown — he was Robert's boss, the President of Caltech. Immediately Harold Brown graciously offered to host our wedding at his home. I had planned a simple wedding ceremony with only two witnesses at the Cathedral of Ulm in southern Germany. This was in reaction to the huge wedding that my parents had had, which had ended up in a bitterly failed marriage. But when the President and

Robert and Juliana Christy with their wedding hosts Harold and Colene Brown, the President and First Lady of Caltech

First Lady of Caltech went so far out of their way, Robert and I felt that we couldn't say no to their gracious offer.

The wedding was on August 4, 1973 because that was the month when Robert as Provost had the fewest responsibilities.

Robert and Juliana at their wedding

Robert and Juliana at their wedding

Robert and Juliana at their wedding

Below is a poem (written by Gerhard Tersteegen) that I gave to Robert on our wedding day. It was originally in German but I translated it into English:

For Robert

I pray to the power of love
Which can reveal itself through us
I will offer myself to the opening buds
By which I have already felt loved
I would like instead of thinking about myself
To immerse ourselves into a sea of love

How much you have already been balancing me
How much your heart has called out to me
By love I am gently and strongly pulled to you
My whole being bows to you
You, kind Robert, you of nature good
You now have me, as I have chosen you

I feel you are he whom I must have
And I feel I must be just for you
Not just in flesh, not just with gifts,
But in your soul my place of rest will be
Therefore I will follow you in all your ways.

First day of married life: at a park on August 5, 1973

For our honeymoon we chose to spend three weeks exploring Greece and another week in Germany. We combined our honeymoon with an International Astronomical Union meeting in Warsaw, Poland, where I gave several invited talks.

Warsaw was not far from the region where I was born (in what used to be Prussia) and from which some of my ancestors had originated. I found the area to be incredibly beautiful and felt as if I belonged there.

Despite our trying romance prior to our marriage, it turned out to be a very happy marriage since all the rough spots had been cleared up

Above and below: Juliana and Robert on honeymoon in Greece

beforehand. The dancing that had started in 1968 in Toronto continued in the lanai and in the living room of our new home. Being together in the kitchen while doing something as ordinary as cooking became an exciting event. To plan and plant a new garden together became another fulfilling adventure.

Swimming, body-surfing, hiking, and skiing all continued, with a new happiness now that the stress and uncertainty were no longer present. The two of us also started to spend time horseback riding together, first on horses rented for five dollars an hour from the Equestrian Center of Los Angeles, and continuing thereafter over the years at the Hunewill Ranch behind Yosemite. This was a dude ranch which had more than a hundred

horses to accommodate riders of every skill level, with hundreds of cattle and many miles of meadows and creeks.

The two of us would spend at least a week at Hunewill every summer, riding at least six hours a day. Robert was an excellent rider in spite of never having had any formal instruction. The two of us would gallop along holding

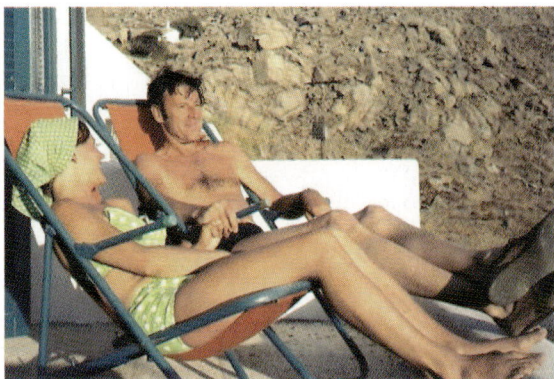
Robert and Juliana on honeymoon in Greece

hands. which required that our horses be in perfect rhythm with each other. When we encountered a six-foot-wide creek, I would let go, slow down, and trot through the water, but Robert would always continue at a gallop and jump the creek, never falling off his horse. For years Robert rode a horse that had been named Rocket because of its ability to propel itself over those wide creeks. This riding continued through Robert's 70's and 80's, and into his 90's.

Robert on the first riding vacation at the Hunewill Ranch

Juliana on the first riding vacation at the Hunewill Ranch

One of the points of our verbal marriage contract was that Robert would be able to enjoy his life and not be limited in his activities if we had a child. When the daughter I had dreamt of, Ilia, was born, I insisted that he finally get a sports car as our family car, something that he had dreamt of all his life — especially after being allowed to drive Oppenheimer's sports car from Berkeley to Caltech when he was a graduate student. Robert chose a green Datsun 280Z.

When we went for a week's vacation skiing or riding, it turned out to be impossible to fit all the baby essentials plus our own skiing or riding gear into this small sports car. However, one time while driving on the freeway Robert noticed another car pulling a very low, broad trailer that would not limit the rear-view visibility in a low-slung sports car. Robert investigated such trailers and found that this low-profile trailer was custom-made. He ordered one, and in the following years this enabled us to fit everything in when going skiing or riding for a week, even when we had not one but two babies plus a nanny (who squeezed into the tiny seat in the back of the sports car). Later when we bought a mountain valley and Robert wished to construct a road along the steep slopes of one side (so that the public would not be driving through the center of the valley), he used this sports car to test the drivability of the road he constructed. Robert loved this sports car and drove it for nearly two decades, until it had over 200,000 miles on it.

Robert was an enthusiastic dancer. His first interaction with me had been dancing together in a psychedelic basement student hang-out at the University of Toronto. We had danced in Portugal, in Switzerland, in Austria,

Robert, Juliana, and first daughter Ilia at age 4 months in 1974

and in Germany. When I had visited him for the first time in Pasadena, we had danced in his living room, and had travelled across town to go dancing in Beverly Hills. After we married, the dancing continued in our new home. We even took dancing classes at the Pasadena City College to acquire more style. Later Robert danced with our daughters, sharing his enthusiasm with them.

Robert loved poetry, and often in daily life he would recite poems that were appropriate that he knew by heart. One of his favorite poets was Robert W. Service, who has often been called "the Bard of the Yukon," and he enjoyed quoting from poems such as "The Cremation of Sam McGee" and "The Shooting of Dan McGrew." He was fond of quoting from Samuel Coleridge's "The Rime of the Ancient Mariner." He also liked the Irish poet Yeats. At the Hunewill dude ranch where we spent a week or two each year, there was a talent night each Thursday evening attended by about fifty guests and several dozen staff. Many poems were written and recited for these evenings, and on at least one occasion Robert recited a poem there.

As an expression of his feeling towards me, on my birthday in 2012 he quoted to me:

"A violet by a mossy stone
Half hidden from the eye!
—Fair as a star, when only one
Is shining in the sky."

from the poem "She Dwelt Among the Untrodden Ways" by William Wordsworth. Robert also had fun writing a few poems. He deplored the fact that children today don't learn enough poetry in school.

Robert loved to sing and to lead others in singing. He had a full resonant voice which encouraged everyone else to sing along with him. He loved it when I listened to his singing or humming. He enjoyed singing folk songs vigorously while accompanying himself on the piano. He would sing while driving in the car, and sometimes in the evening at home, sometimes when colleagues were visiting, but most of the time just with the family. These would be fireside folksongs, such as "I've Been Working on the Railroad," "There Is a Tavern in the Town," "She'll Be Coming 'Round the Mountain," "Oh Susannah," as well as sea chanteys. At Christmas he always liked to sing carols such as "Good King Wenceslaus," "God Rest Ye Merry Gentleman," "The Twelve Days of Christmas," "The First Noel," "Good Christian Men Rejoice," "Deck the Halls," "Joy to the World," and "Hark the Herald Angels Sing," as well as German Christmas songs such as *"Alle Jahre Wieder," "Oh Tannenbaum,"* and *"Oh Du Fröhliche."*

Robert and Juliana: the fourth wedding anniversary, in 1977

We were both alike in that we enjoyed fresh flowers in daily life— in the dining area, in the bedrooms, in the bathrooms. Robert would often go to the most exquisite florist in town. He would go into their large refrigerated storage room, even when he was in a wheelchair, to pick out meaningful and exquisite flowers. He would give these to me on my birthday, our wedding anniversary, Valentine's Day, and other special occasions.

In later years several dinners were given by Caltech to honor Robert on various occasions. There was one occasion when a major dinner was given at the Athenaeum to honor him, with all of the faculty and their spouses invited. There were many speeches. At one point Robert spoke. He expressed how much Caltech had meant in his life, and how he had turned down many offers of positions elsewhere, but finished by saying, "The best thing that ever happened to me was Juliana."

After nearly four decades of marriage, when I was in the hospital recovering from a major surgery, our younger daughter Alexa wrote, "It has been a pleasure to witness my parents sitting cheerfully together and holding hands while napping in their respective chairs, both dealing with health challenges and just happy to be together. My father looks for every opportunity to come to the hospital and just be nearby." When I was recovering at home I had to sleep in a bedroom some distance from the master bedroom. Robert would frequently struggle to visit me and kiss me goodnight, despite his poor vision and the fact that he was trailing over fifty feet of oxygen tubing behind him.

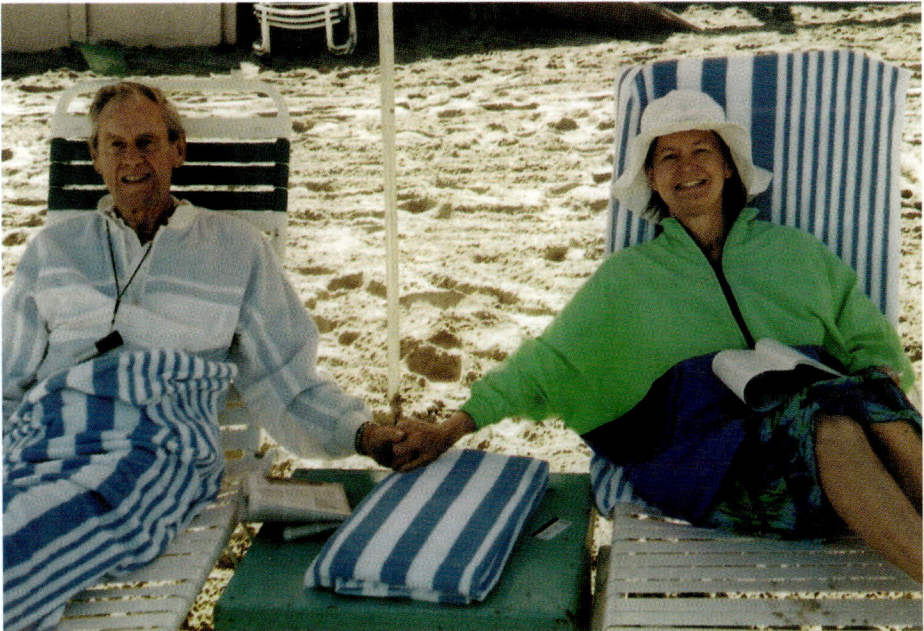

Robert (age 94) and Juliana at the La Jolla Beach and Tennis Club in 2010

Chapter 13

Daughters Late in Life

Robert and I married when he was 57 years old and I was 31. Our first daughter Ilia Juliana Lilly Christy was born a year later, when Robert was 58. Our second daughter Alexandra Roberta Christy ("Alexa") was born two years after that, when Robert was 60. Even though he had been so reluctant to have children with me, Robert turned out to be the

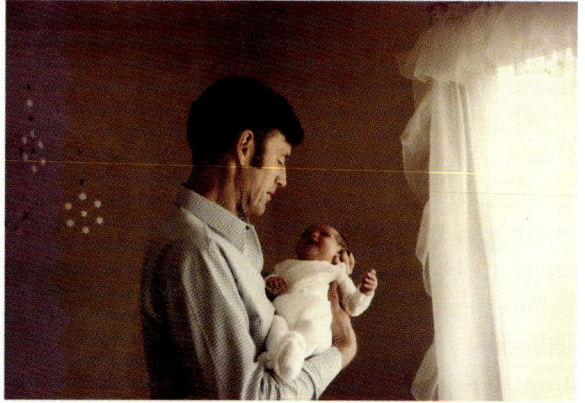
Robert with Ilia, age 3 days

best father imaginable — perhaps because he knew that I could and would take care of them, and thus his interactions with them were a joy rather than an obligation. In Robert's words, it was an "interesting new experience" for him to have daughters to deal

Robert with Alexa, age 4 months

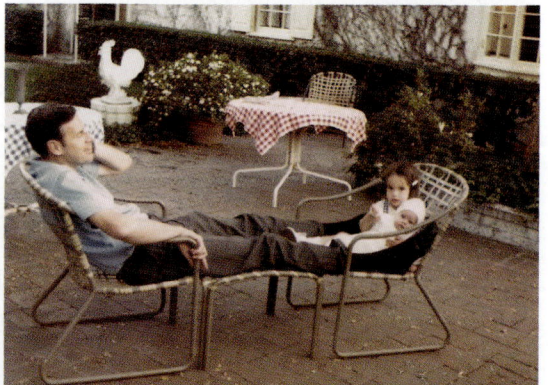
Robert with toddlers Ilia and Alexa: on a "boot ride"

236

Juliana, Robert, and Ilia at age 4 months

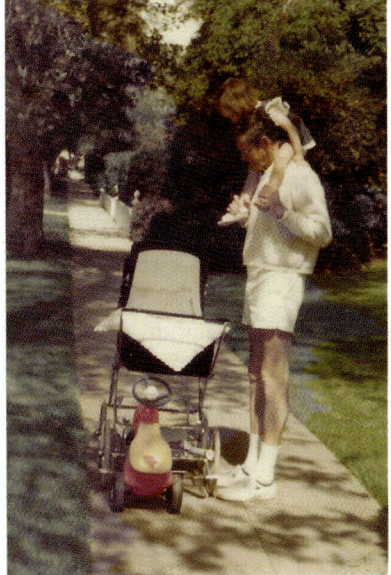

Robert walking with Ilia on his shoulders and Alexa in the pram

with. When they were babies, he was well known in the neighborhood for pushing the big baby pram. At Caltech he was often

Ilia at age 4 (at right) and Alexa at age 2, at the Palm Springs Trustee Meeting in 1978, where they were very popular with the Caltech Trustees

known to have a toddler on his shoulder when visiting the student houses. One year Ilia was so frequently seen at the student houses that her picture was included in the student newspaper. As babies and toddlers, the girls came to the weekend-long Trustee meetings, and were very popular the Caltech Trustees, including Tom Watson, Bill Keck, Shirley Hufstedler, and others.

Both Ilia and Alexa fondly remember a game that Robert played with them called "Big Hug, Little Hug." Alexa recalls, "When I was very young, Daddy and I would play a game I had invented where I would sit facing him on his lap and request a small hug, medium hug, or big hug, and he would enthusiastically comply with the best hugs you can imagine. He would spend considerable amount of time humoring me and playing along as though there was nothing else he would rather do." Ilia remembers him laughing and playing this game with her as well.

Robert's unusually long legs made a strong impression on his daughters. Ilia says, "He was so tall. I was so delighted when I was finally as tall as his legs. He walked with a long stride and I would have to really focus to keep up with his long legs."

Just as with his sons, Robert loved to read to his daughters, and did so most evenings. Ilia recalls, "In the early years, I remember how he would sit with us in the evening while my mother was preparing dinner, and read to us with Alexa on one knee and me on the other. He

Ilia at age 2 with her father: not yet as tall as his legs!

read through the entire 'Little House on the Prairie' series of books this way." He also read many other books to them this way, including "All Creatures Great and Small" by James Herriot, which sparked Ilia's interest in veterinary medicine. This later evolved into a career in medicine. She pursued a dual specialty in pediatrics and internal medicine.

Robert treated his daughters much like his sons: they learned to change tires and construct toy cars. Ilia began learning to ski when she was four years old, and Alexa when she was two. Robert taught the girls how to ride bicycles. He taught them an appreciation for food and cooking. Ilia was a born dancer, and Robert had great fun dancing with her when she was young. Alexa caught on to the fun of dancing too, and Robert would dance with both of them.

There were four languages in daily life in our home. Robert was responsible for English, I spoke German, household help spoke Spanish, and Caltech Chinese students and other Chinese teachers taught our daughters Mandarin Chinese.

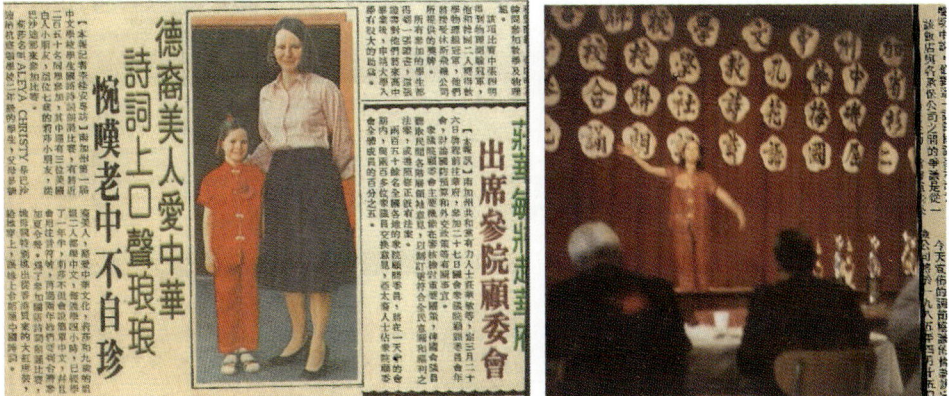

Above left: Alexa at age 6 wins an award in a Chinese recitation competition among 22 schools for Chinese children. Above right: Ilia at age 9 wins an award in a similar Chinese competition.

Both Robert and I spent a great deal of time instilling in our daughters a love for learning, turning math and physics into games. Eventually, when the girls left to go to college at Stanford University, Robert missed them greatly, and continued having fun doing calculus and physics with them over the phone. Ilia once described this:

> He had a tremendous curiosity and loved solving problems. He was patient and always helped me with my homework when I needed it. When I was in third grade I had trouble learning my multiplication tables. I was embarrassed. My father was brilliant and yet I was struggling with 3 x 3. He wrote out a grid for me so that I could practice again and again. And eventually I got it. As I advanced in my math education he had more and more fun. He loved hearing my challenging calculus problems and figuring them out on his own. His favorite time was when I was at Stanford. I would sometimes call him in the evening with a physics or calculus problem I was stuck on — and he would ponder it. By morning I had a phone call with a clear explanation of how to solve it. I had to bring home all my exams so that he could also work through those problems for fun.

Alexa recalls,

> Daddy was an excellent teacher because he was incredibly selfless, patient, and encouraging. For instance he would ride, ski, and play tennis to our level as children despite being quite proficient himself. He never appeared bored and was so patient, kind, and encouraging while coaching us. Daddy always gave us the impression he would rather do these activities with us than with his peers. In time his teaching paid off and we were able to keep up better and I have so many wonderful memories of enjoying these activities with him.
>
> Of course Daddy also taught us academics by persistently yet gently correcting our grammar and by always helping us with our homework. Even in

junior high when the material was so basic it must have been mind numbing he was always interested, available, helpful, encouraging, and never condescending. When the content became more complex he was excellent at explaining everything and making it as simple and clear as possible. When I did not have much talent or passion for physics, Daddy was understanding and not at all

Robert teaching tennis to Ilia and Alexa (ages 8 and 6)

disappointed. Furthermore he enthusiastically supported my choice to become an engineer. He still helped me with my homework even in subjects with which he was less familiar because he loved learning as well as teaching.

Daddy also taught us how to drive, haul trailers, use hand tools, assemble furniture, cook Thanksgiving dinner, change a tire, and do so many other countless things with the same kindness, patience, and encouragement that so effectively built not only our practical skills but more importantly our confidence and trust in ourselves. He always believed in us and gave us the impression we were capable of anything. I know that what we have accomplished in life is largely due to his influence.

Skiing, Horses, Gymnastics, Tennis, and Polo

It was part of my verbal marriage contract with Robert that having children should not impede his opportunities to play the sports that he loved. Even when the children were only toddlers we took them along on our skiing trips, and left it up to them whether they wished to remain inside or be involved in the skiing outside. They always chose the latter, and started learning to ski while very young, when Ilia was four and Alexa was

Ilia trying on ski boots at age 1 ¼ (though she did not actually start skiing until age 4)

Alexa skiing at age 2: at left she is drinking from her milk bottle, at right she is pushing along

two. Alexa remembers, "One day when I was very young, Daddy fearlessly jumped off an ascending ski lift when he noticed I had not made it on. I thought he was a superhero just leaping off to be by my side and encourage me to try again with the next chair."

As soon as she could, Alexa started to do moves that might have been expected of someone with gymnastic training; as soon as she could walk, she coaxed Robert into trying to do cartwheels and headstands with her. When Alexa was two and Ilia was four, they started gymnastics classes.

Alexa the gymnast in the garden at age 5

Alexa was fascinated by gymnastics, and practiced a great deal wherever she could, not just in the gym but also in the garden and in the house. Robert would coach her, and use his understanding of physics to correct the form of her back-flips. When Ilia and Alexa started to ride in equestrian jumping competitions, they constructed wooden horse-jumps for people in the back yard, and Robert showed them how to do the carpentry. Later, Alexa joined the track team at

Ilia on a horse at age 1 ¾ — drinking from her bottle

her school, and Robert again coached her on her jumping form and her sprinting.

A love for horses was evident in our daughters at an early age. When Robert and I went on a week's vacation intending to ride every day, we ended up walking and running while leading a horse because our

Alexa riding at age 2 ½

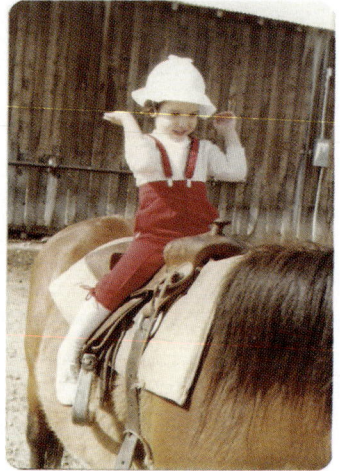

Ilia happily riding at age 2 ¾

1¾-year-old daughter Ilia insisted on riding too. When Alexa was only six months old she saw her first horse, and was not happy until she was on top of it. This developed into a serious involvement with horses, the girls having formal riding instruction for the next two decades.

Ilia and Alexa competed in many riding events. Robert would drive them there, often at 5:00 a.m., and would help them to prepare for the competition. In the evening he would help care for the horses, including unbraiding the horses' manes (which he could easily do, being so tall). These competitions were initially local, but as time went on and the

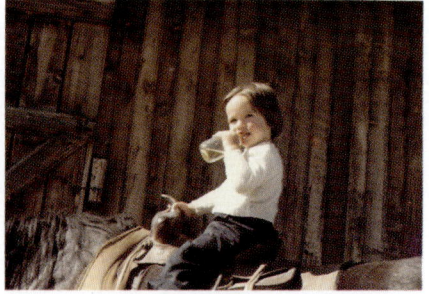

Alexa re-invents gymnastics on horseback at age 3

girls' skills increased the competitions become state-wide. Both Ilia and Alexa became very skilled and accomplished riders.

When Ilia was admitted to Stanford, which had an equestrian team, Robert persuaded Ilia to learn horse polo rather than continue with horse jumping. Polo was a sport that he would have liked to do if he had had the opportunity when he was younger. Ilia became increasingly skilled at polo, spending 40 hours a week on it, and eventually was named the top female intercollegiate polo player of the U.S.A. in 1997. Robert was so tempted by polo that he even started polo lessons himself in his late 70's. It was hard to figure out when Ilia could have had the time to study, but she

Robert with Ilia and Alexa

nonetheless managed to get all A's and B's at Stanford University.

Alexa's love was vaulting, a form of gymnastics on galloping horses. She tried to get Robert into that, and there were many hilarious episodes of him trying gymnastic cartwheels and headstands on the bedroom floor — these attempts did not go too well. At one point on a moonlit night while his horse Blueberry Pie was eating, Robert climbed onto the feeder and tried to get onto the horse as he had seen Ilia and Alexa do. But the horse was startled and jumped away. Robert slid off awkwardly, and broke a rib that took seven weeks to heal. Another time, when the family was galloping down a dirt path, Robert's horse bucked and the two parted company. Despite his 6'5" height he managed to somersault in the air. He stood up and bowed graciously to us as if it had been a stunt that he had meant to perform.

Ilia once described her father's enjoyment of horses, and his sharing of that enjoyment with her:

> He and my mother loved horseback riding and shared this enthusiasm with us. Because they wanted us to learn to ride and to keep up with them, they offered us riding lessons and supported our passion for horses as it grew. However he thought that the subjective nature of judging at horse shows was somewhat arbitrary. He preferred sports with more danger and excitement. When we competed at timed jumping events he proudly cheered us on. His favorite times though were when we played polo at Stanford. Finally we had found an equestrian discipline that was thrilling and competitive. Polo season saw my parents spending most weekends on the road so they could attend our games. It was not enough though to watch his girls from the sidelines. With his competitive spirit still strong, he learned to play polo in his late 70's.

Travels Near and Far

In the mid-1980's Robert and I, along with our daughters, travelled to the Los Alamos area that Robert loved so much. We started out at a dude ranch near Santa Fe, exploring the strikingly beautiful mesas and valleys around Los Alamos as well as the nearby caves and Indian pueblos. We then continued

Robert with Ilia

our exploration of U.S. history by travelling across the country to Jamestown, Virginia, where the early European colonists had first landed. We included visits to several early American towns, and made a special point to study the homes of the first American presidents, including George Washington and Thomas Jefferson. The Washington and Jefferson houses influenced some of the features of the house that we later built in our own mountain valley, a ranch 90 minutes north of Caltech.

In the late 1980's, Robert and I took our daughters to Germany and Austria. We visited my good friend Heidi Hering, with whom I had kept in touch since my childhood days. I introduced my family to the Wingerters, the wonderful couple in

Robert with Alexa

Munich with whom I had stayed for several years while I was working at the Max Planck Institute. For our daughters, though, one of the highlights of the trip was visiting the Lipizzaner horses in Vienna and the Austrian countryside where these horses were bred and raised.

In the 1990's, the family made two restful and enjoyable week-long trips to Hawaii (though on one of these Ilia was unable to come because she was at Stanford). We all learned to windsurf. Some could remain standing and sailing for longer than others. Robert could only manage to stay up for one or two seconds, but Alexa became very good at it. She awed the rest of us with her skill.

In 1997, I tore the ACL (central knee ligament) in my right knee in a horse riding mishap. Robert had just lost the central vision in both eyes due to macular degeneration though his peripheral vision was still good. One could say that the two of us had become "the lame and the blind," respectively. So to raise our spirits, and demonstrate that these problems were not truly crippling, we decided to hike in the Himalayan Mountains of Nepal (near Mount Everest), accompanying my sister Christel and her husband Erich who were enthusiastic and seasoned travelers. At this time Ilia wished to travel around the world. Since none of her friends were able to go with her she decided to join the trip to Nepal, and then to continue around the world on her own. The five of us set out on this trip and spent two weeks in Nepal, going hiking in the mountains of the Annapurna Conservation Area. Even though Robert and I were both significantly handicapped, we had a great deal of fun. Ilia describes this trip:

> He [Robert] is the best travel partner I know. When he was in his 80's we trekked in the Annapurna Mountains in Nepal together. Despite the fact that he had very little vision, he enthusiastically hiked for days along those steep mountain trails, taking in the incredible scenery and cherishing being immersed in stunning natural beauty. Throughout his life he so enjoyed being in the mountains. He loved the adventure and exploration of travel. The hardships and inconveniences of being in rural and basic conditions with no running water and non-western toilet facilities did not bother him. He never complained. I always said I could travel anywhere in the world with him.

In 1999, Robert and I travelled with our daughters to China and Pakistan for six and a half weeks. When the children were young they had been promised that if they learned proper Chinese the family would all go to China together one day. Robert had always wanted to travel along the Silk Road, the ancient trading route across central Asia. We planned to start on the east coast of China and to fly from one city to the next where local drivers and local translators were made

Robert at age 81 (center), Ilia (at right), Alexa, Juliana, and Juliana's mother Lilly Stelter (at left)

available to us. We travelled from Beijing and Xi'an to the border of Tibet, into the desert area around Urumqi, and on to the Central Asian trading city of Kashgar. From there we drove through the Karakorum Mountains bordering China, Pakistan, and India. It was a strenuous but illuminating trip. Robert was very interested in seeing some of the highest mountains in the world, along the Karakorum highway in Pakistan. We visited villages of the Hunza people, which was one of the key attractions for me. We then travelled through the mountains to Gilgit, and on to Islamabad and Lahore in Pakistan. I made a side trip to Karachi — just myself and our guide — to buy some beautiful oriental rugs. Throughout Pakistan we were treated with great hospitality, and they frequently said, "Please have more of your American friends visit us in Pakistan."

In 2006, when Robert was 90 years old, Ilia organized a family trip to Santa Fe and Los Alamos as a family reunion and to hear his stories of his work there during the Second World War. Ted and Peter and their families joined us on this trip. A white water rafting trip was organized by the more adventurous, with Robert enthusiastically taking part with his children and grandchildren. Not many 90-year-olds go white water rafting.

Robert aged 90 (center at back) on a white water rafting excursion near Los Alamos in 2006

The Daughters' Professions

Although Robert encouraged a love of learning in his daughters, he did not try to push them into any specific profession, instead supporting them in their own goals. Ilia says,

> He was incredibly supportive. When I was enthusiastic about something, he wanted me to follow my passion. Alexa and I wanted to work at the Hunewill Ranch as assistant wranglers in the summer and he encouraged us — even though we were only 13 and 15 years old. He did not expect me to follow him in physics, but instead wanted me to find what it was that sparked my curiosity, and to pursue that — which turned out to be biology and then medicine.

Ilia obtained a bachelors degree from Stanford in biology. After the trip to China and Pakistan, she entered medical school at the University of Rochester, and earned an M.D. in 2002. She went on to specialize in internal medicine and pediatrics at the University of California at San Diego. She is presently (2013) a physician in Internal Medicine at the Scripps Clinic in San Diego.

Alexa obtained bachelors and masters degrees in industrial engineering from Stanford. She is presently self-employed, running her own business.

When Robert was in his late 80's and 90's and had multiple hospitalizations and surgeries, Ilia and Alexa would come,

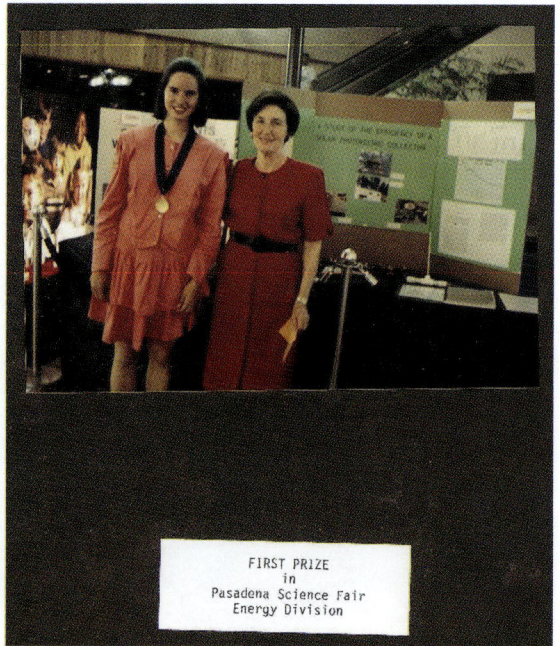

FIRST PRIZE
in
Pasadena Science Fair
Energy Division

Ilia with her tenth-grade science teacher, Roza Kupperman, who sponsored Ilia for her prize-winning project, "A Study of the Efficiency of a Solar Photovoltaic Collector" — though it was Alexa who ended up studying engineering in college

sometimes alternating being with him day and night to watch over him and check for mistakes by the hospital staff. Ilia took a leave from her residency program at one point, to be with her father at critical times. Alexa insisted on driving and accompanying her father to his many doctor's appointments, all over the Los Angeles area. When away from him, they telephoned him almost daily to connect with him.

Robert's life was enriched by the activities that he shared with them and by their care for him. They helped to keep him young. As with his sons, he was very proud of what Ilia and Alexa have accomplished.

Illustrations of the Equestrian Accomplishments of Robert's Daughters

Above left: Ilia with a first place award. Above right: Ilia jumping her own horse Addie.

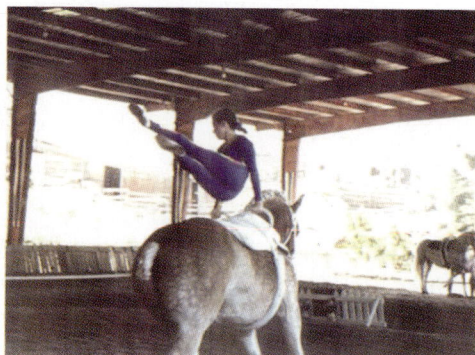

Above left: Alexa jumping her own horse Likely. Above right: Alexa in a 1990 national vaulting competition.

Above left: Alexa and Ilia on the Stanford polo team. Above right: Ilia and Alexa recovering from injuries received during their equestrian competitions

In 1997, Ilia was named the female Intercollegiate Player of the Year. Article reproduced courtesy of the Pasadena Star-News.

Staff photo by KEITH DURFLINGER

STANFORD UNIVERSITY polo team member Ilia Christy of Pasadena holds the female Intercollegiate Player of the Year award she received recently in Oklahoma.

Intense commitment to her college polo team led to a Pasadenan's status as best woman player in the country

ILIA CHRISTY takes a shot while playing for Stanford University.

By Janette Williams
STAFF WRITER

PASADENA — With its jet-set overtones of privilege and big money, polo doesn't come to mind as your typical college sport.

But for 23-year-old Ilia Christy of Pasadena, just named top woman college player in the country, it's been polo on a shoestring for the four years she's played for Stanford University.

The team constantly scrambles for the $50,000 a year it takes to keep the program going using donated polo ponies and sharing equipment — even shirts — with the men's team.

But none of that stopped Christy, who had never even played the game before college.

"She's the best player I've ever had," said Wes Linfoot, the Stanford polo coach. "I don't know how she kept her grades up and dedicated that much time to polo. Personally, I have no idea how she could do that."

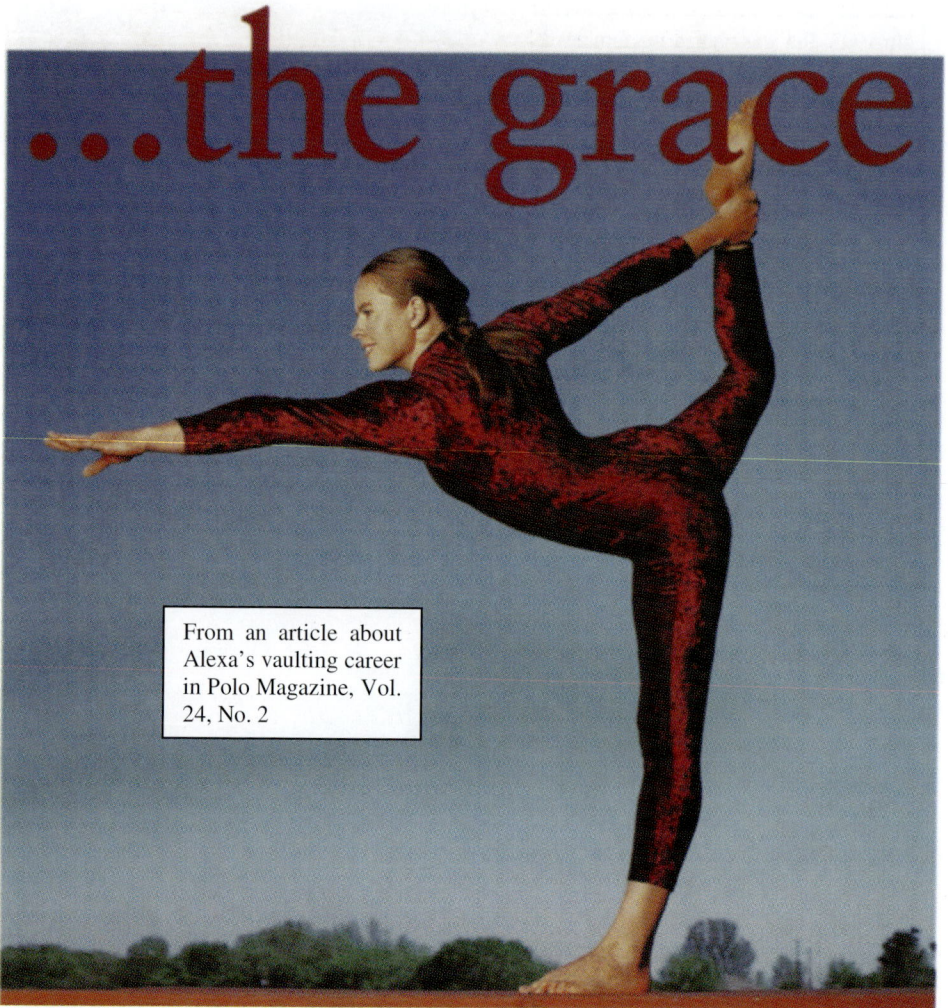

From an article about Alexa's vaulting career in Polo Magazine, Vol. 24, No. 2

Alexandra Christy
STANFORD UNIVERSITY

In Hungary and Germany, horse vaulting is almost as popular as bike riding is in America. "Every kid wants to do it over there," says Stanford senior Alexandra Christy of the sport that is often described as physics on horseback, so seemingly cerebral are its airborne maneuvers. No wonder, then, that the half-German daughter of two brilliant physicists took to it so naturally, catapulting herself to a national title at the bronze level of competition in 1991. When she zeroes in on a target, the 21-year-old,

Pasadena-born industrial engineering major has the portentous accuracy of a heat-seeking missile, according to her teammates, who describe their co-captain as "intensely focused." While recovering from a vaulting injury that short-circuited her competition schedule three years ago, Christy rerouted her gymnastic urges to the Stanford Polo Club's hitting cage, where she aimed to match the legendary polo skills of her sister, a former Stanford All-American. ◆ *"She didn't exactly have to twist my arm," says the elegantly elastic Alexandra of sister Ilia, who persuaded her younger sibling and ranch partner (the family is building a ranch in the California hills with space for retired polo ponies) to go for the goal instead of the gold last year at the West Coast Regional Championships. Not surprisingly, Christy felt right at home on the back of a polo pony. She was named one of the top eight players in the country and also earned All-West All-Star honors.*

POLO MAGAZINE **50** JUL/AUG 1998

Chapter 14

Interim President of Caltech

In 1976 with the U.S. Presidential election campaign underway, Caltech's President Harold Brown was spending most of his time in Washington, D.C. Robert filled in for him as Acting President during this period. Not long after Jimmy Carter was elected, Harold Brown left Caltech to become the U.S. Secretary of Defense. In January 1977 the Trustees appointed Robert Christy as the Interim President of Caltech. However, he also retained his position as Provost. He therefore had to keep up with two demanding full-time positions simultaneously.

(Caltech Photo)

Fall meeting of the Caltech Trustees, at Smoke Tree Ranch, Oct. 1977. <u>Back row:</u> J.E. Robison, H.J. Haynes, B.F. Biaggini, R.W. Galvin, L.R. Wasserman, S. Ramo, F.G. Larkin, R.S. McNamara, and S.D. Bechtel, Jr. <u>Middle row:</u> W.M. Keck, Jr., M.L. Scranton, S.R. Rawn, A.B. Kinzel, H.G. Vesper, T.J. Watson, Jr., R. Anderson, and W. Burke. <u>Front row:</u> G.W. Fitzhugh, H.J. Volk, F.L. Hartley, R.F. Christy & R.S. Avery standing in front, S.M. Hufstedler, C.J. Medbury III, A.O. Beckman, and E.M. Jorgensen. (Caltech photo)

Caltech Associates hear Anderson

Caltech's Associates number 585 and more than 450 of them turned out for a dinner party at the Beverly Wilshire Hotel this week where acting president Dr. Robert Christy had the fortuitous opportunity to tell them about

AT CALTECH ASSOCIATES' DINNER — Dr. Robert F. Christy, left, acting president of Caltech, greets evening's speaker, Robert O. Anderson. In background is Joseph B. Earl, president of the Associates.

—Staff photo by Ed Norgord

ENERGETIC DISCUSSION — Gov. Edmund G. Brown Jr., left, in Pasadena to discuss the need for coal fired plants to generate electricity to meet the future needs of California, talks with Dr. Robert F. Christy, acting president of Caltech, center, and Dr. G. Alex Mills, U.S. Department of Energy.

The above photos and captions are reproduced courtesy of the Pasadena Star-News.

Above and below: Robert Christy as Interim President of Caltech, making presentations (Caltech photos).

Robert was an exemplary President for Caltech, always putting Caltech's interest first and his own needs second. He was very much admired by the Trustees, and by

everyone who worked with him. He was totally involved, an outstanding administrator who was good with people and clearly capable of handling any problems that arose.

Removing a Road that Divided Caltech

San Pasqual Street was a fairly heavily travelled east-west public road north of and parallel to the very busy California Blvd. San Pasqual Street ran right through the Caltech campus, dividing it in half. Not only was this disruptive, it was dangerous to students and faculty who were trying to get to their classes and laboratories. Robert worked hard to get the City of Pasadena to agree to let Caltech have that portion of San Pasqual Street. In other words, three blocks of a public road were turned into private land. Caltech removed the roadway; in its place are now greenery and pedestrian paths [see the wide orange outline at the center of the map on the following page; the parts of San Pasqual Street that remain are outlined in black].

Robert Christy as Interim President of Caltech, handing out diplomas at commencement (Caltech photo)

Robert Christy with Robert O. Anderson and Lee DuBridge (Caltech photo)

Chairman of the Board of Trustees Stan Avery (at left) and Robert Christy (center) greet a visitor (Caltech photo)

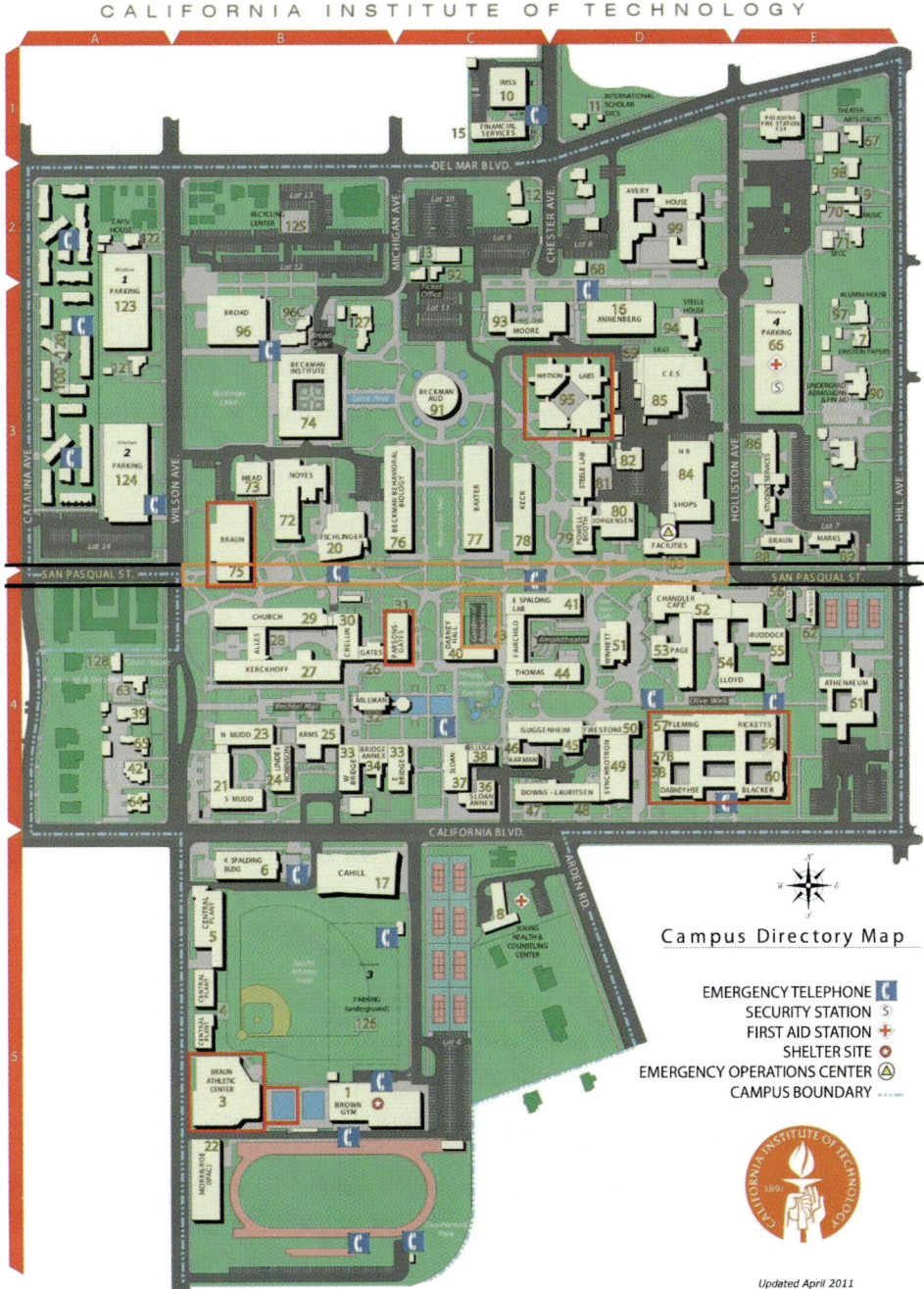

Map of the Caltech campus: red and orange outlines indicate Robert's additions and renovations

Preservation of Caltech's Architecture and Gardens

Garden of the Associates:

Dabney Hall is one of the most beautiful old buildings on the Caltech campus. Attached to it is a lovely garden, the Garden of the Associates, lined with olive trees and enclosed by a high stone wall. Many of Caltech's festivities take place in this garden. For example, in 1983 it was used to celebrate the Nobel Prize that had been awarded to Caltech astrophysicist Willy Fowler. I remember the memorial held there for Max Delbrück, another Nobel Prize winner of Caltech. I remember weddings being celebrated there. But this garden was becoming too small for Caltech's growing faculty. When the adjacent land from San Pasqual Street became available, Robert had the stone wall moved outwards, and the old olive trees replanted further outwards next to the wall [outlined in orange at lower center of the map on the previous page]. This was so well done that today there is absolutely no sign that it had not been this way from the beginning.

The old undergraduate buildings:

The old undergraduate student houses are to the south of the Olive Walk which leads to the Caltech Athenaeum (the beautiful faculty club where many of the faculty enjoy their lunch). These student houses are a set of four interconnected buildings, each built around an internal courtyard. Clockwise from the northwest, they are named Fleming, Dabney, Blacker, and Ricketts Houses. These houses have immense impact on Caltech's undergraduate students; it is sometimes said that Caltech graduates are identified more by the house they lived in than by what subject they majored in. When Robert became part of the administration the finances were not in good shape. Caltech was in the red, with expenditures exceeding income. The old student houses were in disrepair, with leaking roofs and many other problems, and it was seriously considered that they should just be torn down.

On the north side of the Olive Walk there were three new undergraduate student houses, the Page, Lloyd, and Ruddock Houses. These had been built to handle an expanding student population, but they lacked the charm and beauty of the old student houses. Robert struggled to avoid having to tear the lovely old student houses down and build new and less attractive ones. He eventually managed to locate funds for their repair. As a result the four old student houses were preserved and are still in use today. [This set of four linked buildings is outlined in red at lower right on the map two pages previous.]

Parsons-Gates Hall of Administration:

The lovely old administration building, Throop Hall, had been demolished after the 1971 earthquake. The President, Provost, and some other upper-level administrators were inconveniently situated in cramped offices on the third floor of the Millikan library. They had very limited space for their assistants. On the other hand, there was a centrally located building that had been vacated due to earthquake damage. This was a beautiful building that reminded one of the demolished Throop Hall. It was Robert who came up with the plan of retaining the beautiful exterior walls, while the inside was completely dismantled and rebuilt according to modern earthquake standards. The interior was designed to appropriately serve future administrations of Caltech. The Goldberger administration turned out to be the first one to benefit from Robert's foresight. The building is today called the Parsons-Gates Hall of Administration [see red outline at the center of the map two pages previous].

New Buildings for Caltech

I developed a close friendship with Trustee Tom Watson, Jr., the head of IBM. He would always come to the Trustee dinners that I gave in our home as Caltech's First Lady, often flying there in his own plane. He would always send flowers afterwards to show his appreciation of my efforts.

Robert worked hard to persuade Watson to provide funds for a new building. The result was a new building for Caltech, named after Tom Watson's father: the Thomas J. Watson, Sr., Laboratory for Applied Physics. Robert also succeeded in getting funds from Braun, another Caltech Trustee, for the Braun Laboratory for Chemistry and Biology. Robert was crucial in obtaining a top-level new gym for Caltech, the Braun Athletic Center, and a second swimming pool next to this new gym. [These three buildings are outlined in red at upper right, center left, and bottom left in the campus map two pages previous.]

When I first became involved in the weekend-long Trustee meetings at Palm Springs, I was shocked to notice that the intriguing couple, Arnold and Mabel Beckman, were sitting and eating all by themselves with nobody talking with them, looking rather lost. I tried to fill the void that I perceived, and sat and ate with them; it was fascinating to talk with them. They had already contributed immensely to Caltech, including providing the funds for the Beckman Auditorium and the Beckman Laboratories for Behavioral Biology. Every opportunity that I had, I tried to single them out and show them gratitude on the behalf of Caltech. Later, I was very pleased to learn that David

Robert makes a presentation to Olga Taussky Todd, who in 1957 had become Caltech's first woman faculty member (Caltech photo)

Morrisroe continued this close relationship with them and that the Beckmans continued to shower Caltech with their generosity: in later years they provided funds for the Beckman Institute and the Beckman Laboratory for Chemical Synthesis.

In total, Robert's efforts resulted in three new buildings for Caltech, the renovation of the Parsons-Gates administration building and of four student houses, the enlarging of a garden, and the reclaiming of the area in the middle of the campus that used to be San Pasqual Street.

Juliana as Caltech's First Lady

I had been at Caltech for five years and was quite happy in my position, doing my own research. But when Robert became Caltech's Interim President, I interrupted my research and enthusiastically took on the task

Robbie Vogt and his wife Michelene watch (at back center) as Robert makes a presentation to Alan Sweezy (Caltech photo)

of serving as the First Lady of Caltech — there would not have been time to do both jobs.

Among other things, being Caltech's First Lady involved hosting dinners and receptions for Trustees, faculty, and students. With my secretary, I researched each guest's history so as to make sure the right people sat next to each other or at the same table, to avoid having bitter rivals in too long a contact with each other. I put a substantial effort into making sure that the atmosphere set a tone of friendship and relaxation. Each dinner had a different theme, with appropriate decorations, from the flowers to the type of lighting. I never used lighting from electric lights. Instead I had candle-lit dinners without the dilution from electric lights. Sometimes I used candles surrounded by dry ice to produce a diffuse cloudy effect. Sometimes I used glass bowls filled

Juliana and Robert Christy at a Caltech function (Caltech photo)

with water and topped with a thin oil layer, with burning wicks and floating flowers. Outside, I sometimes used torches for light.

For one dinner, which I called the "Bandits' Dinner," my secretary and I carefully slit apart 80 walnuts, removed the nut, inserted 80 different appropriate messages, closed the walnuts again, and carefully placed them inside the napkins. For another dinner, the "Dinner of Magic," I placed recipes for magic potions inside such walnut shells. Images of a place card, an invitation, and a menu for this latter dinner can be seen on the following pages.

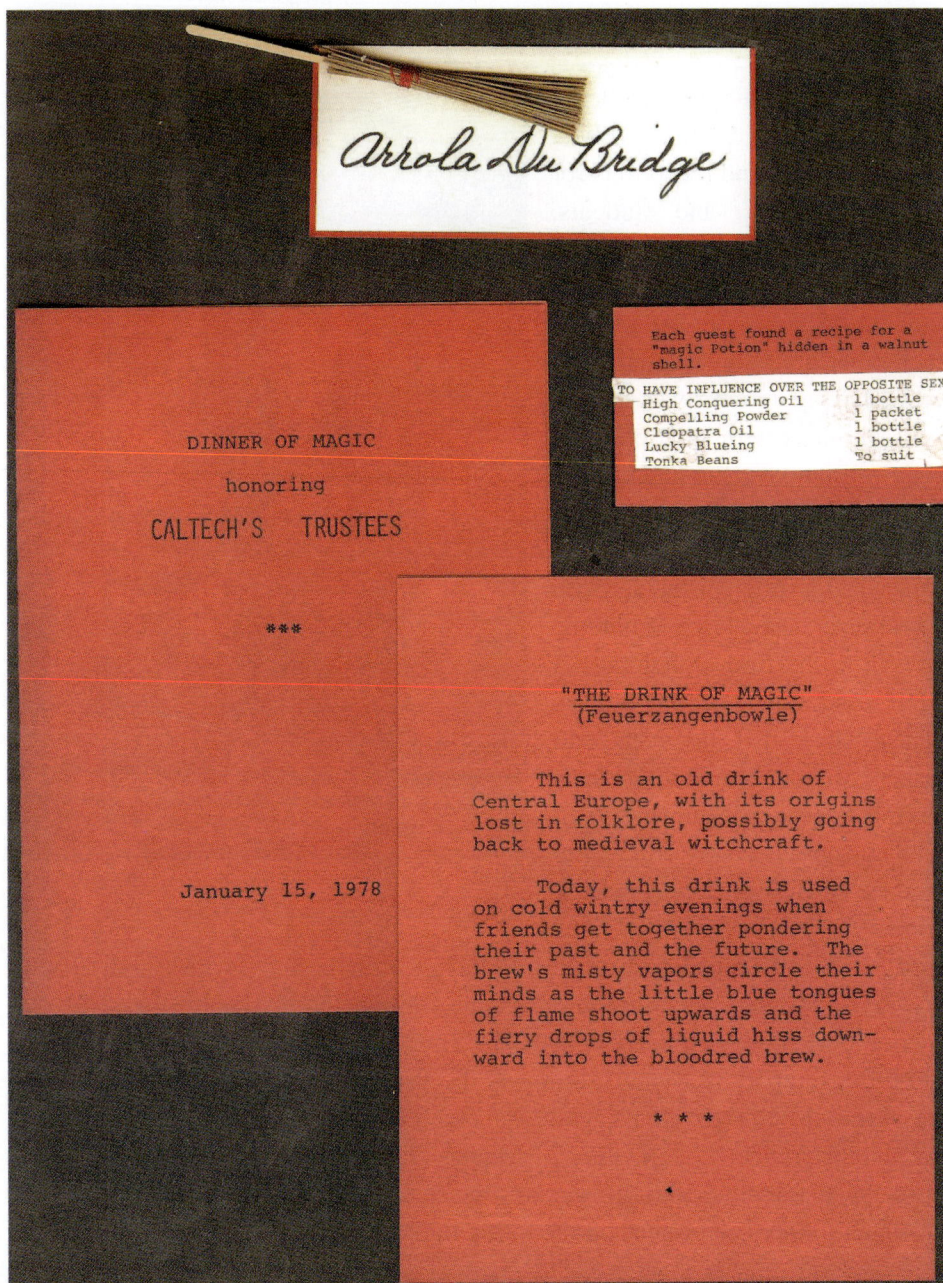

Arrola Du Bridge

DINNER OF MAGIC

honoring

CALTECH'S TRUSTEES

***

January 15, 1978

Each quest found a recipe for a
"magic Potion" hidden in a walnut
shell.

TO HAVE INFLUENCE OVER THE OPPOSITE SEX
High Conquering Oil   1 bottle
Compelling Powder   1 packet
Cleopatra Oil   1 bottle
Lucky Blueing   1 bottle
Tonka Beans   To suit

"THE DRINK OF MAGIC"
(Feuerzangenbowle)

This is an old drink of
Central Europe, with its origins
lost in folklore, possibly going
back to medieval witchcraft.

Today, this drink is used
on cold wintry evenings when
friends get together pondering
their past and the future. The
brew's misty vapors circle their
minds as the little blue tongues
of flame shoot upwards and the
fiery drops of liquid hiss downward into the bloodred brew.

* * *

Items from a place setting for one of the Trustee dinners, the "Dinner of Magic"

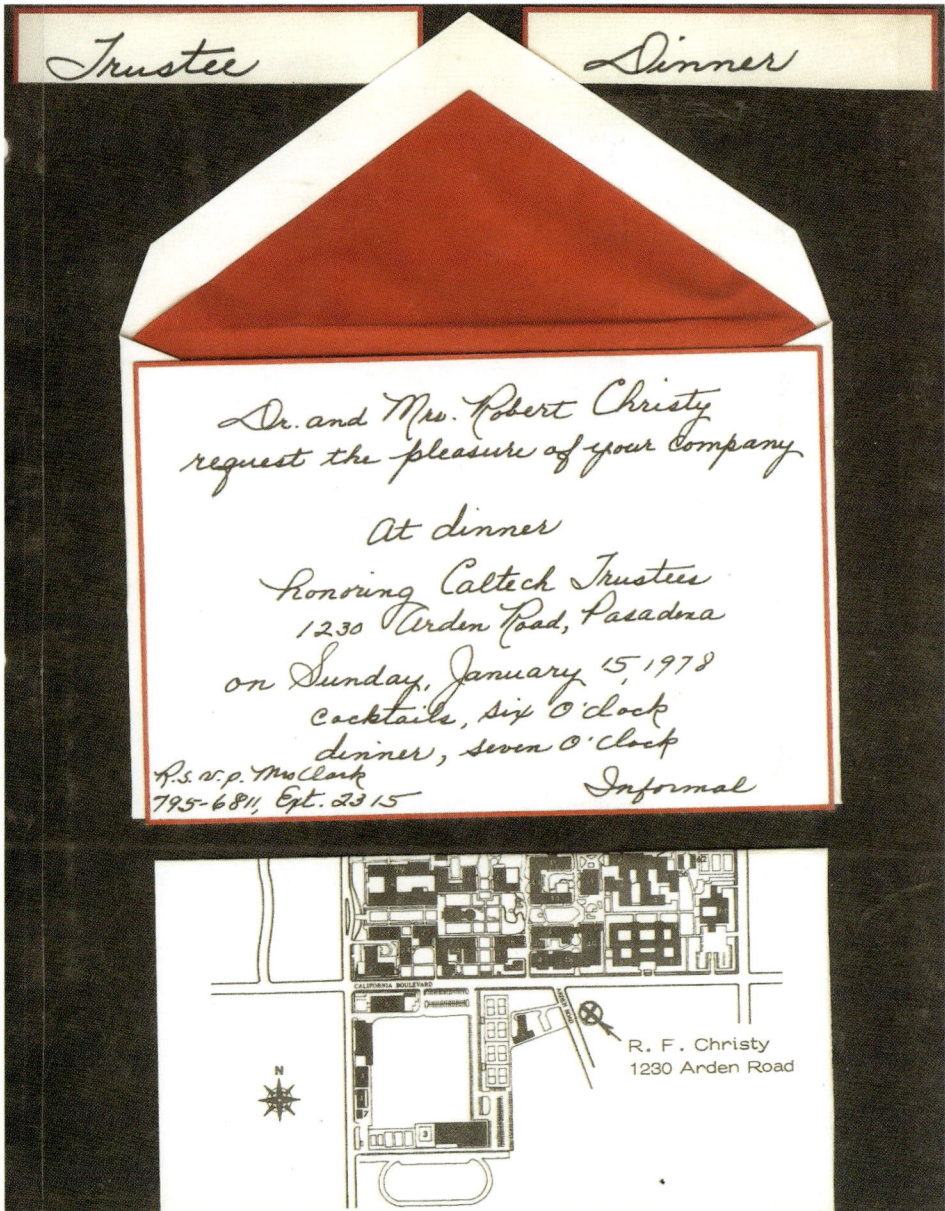

One of the invitations to the "Dinner of Magic"

Just folding the large napkins into fanciful shapes such as flowers, baskets, or boats would take hours. Standard caterer's food was never used. Instead I would practice interesting recipes ahead of time, then find caterers and teach them to cook it in our home as practice, before they cooked the actual dinner for the event (again in our home). I would usually cut flowers from those readily available on the Caltech campus, especially the roses. If not enough were available I would go to a florist to select flowers by the bunch. I would always make the flower arrangements for the centerpieces myself, with the aid of my secretarial aide. I wished to create an intimate personal atmosphere, as well as to save funds for the faculty and students. I never hired professional musicians. The music was provided by talented Caltech faculty, such as the Caltech barbershop quartet, or by talented students. Typically such a Trustee dinner with 70 or 80 sit-down guests in our home would require about ten days of preparation.

I also worked hard on getting the Trustees to meet some of our talented faculty and students at

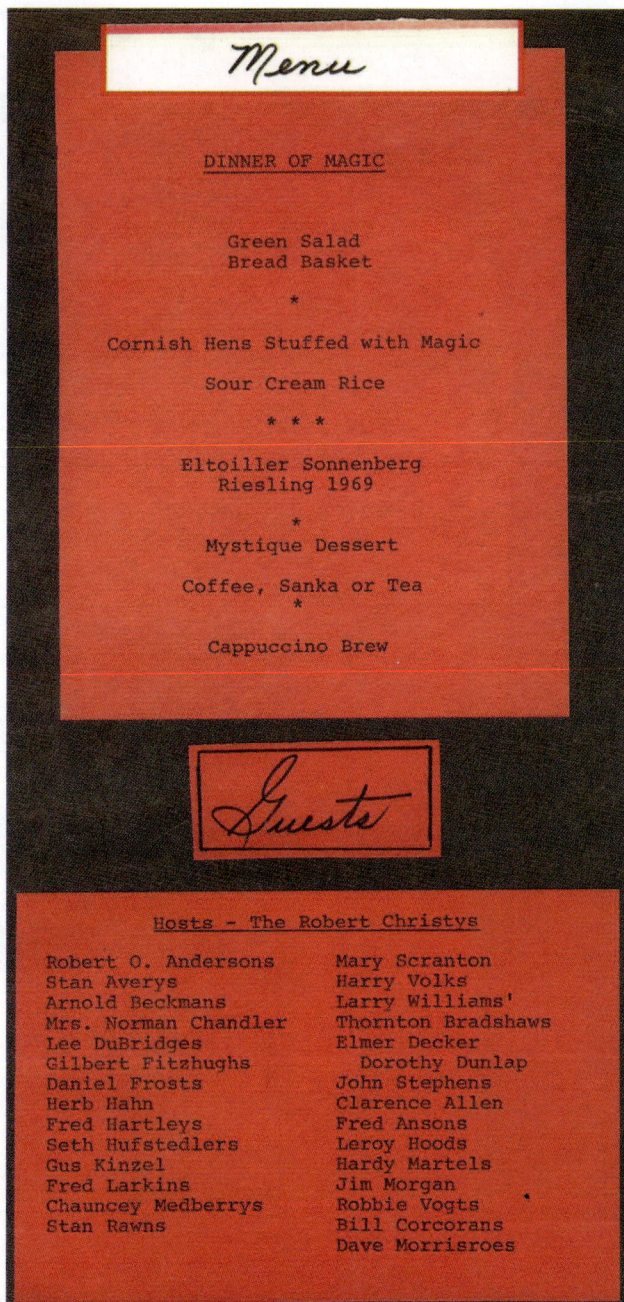

Menu

DINNER OF MAGIC

Green Salad
Bread Basket

*

Cornish Hens Stuffed with Magic

Sour Cream Rice

* * *

Eltoiller Sonnenberg
Riesling 1969

*

Mystique Dessert

Coffee, Sanka or Tea

*

Cappuccino Brew

Guests

Hosts - The Robert Christys

Robert O. Andersons	Mary Scranton
Stan Averys	Harry Volks
Arnold Beckmans	Larry Williams'
Mrs. Norman Chandler	Thornton Bradshaws
Lee DuBridges	Elmer Decker
Gilbert Fitzhughs	Dorothy Dunlap
Daniel Frosts	John Stephens
Herb Hahn	Clarence Allen
Fred Hartleys	Fred Ansons
Seth Hufstedlers	Leroy Hoods
Gus Kinzel	Hardy Martels
Fred Larkins	Jim Morgan
Chauncey Medberrys	Robbie Vogts
Stan Rawns	Bill Corcorans
	Dave Morrisroes

The menu and guest list for the "Dinner of Magic"

ICY CAULDRON — Youngest guest at the party was Alexa Christy, 14 months old. Jessica Frank shows her dry ice cauldron that helped set mood.

MYSTIC BOWL — Arrola DuBridge and her hostess, Juliana Christy, left, peer into mystic bowl. Witchcraft themed dinner gathering for scientific minds.

—Staff photos by Walt Mancini

Above: captioned photos of the "Dinner of Magic" (from a January 25, 1978 article) are reproduced courtesy of the Pasadena Star-News.

Below: article of January 12, 1978 by Jody Jacobs describing the "Dinner of Magic" is reproduced with the permission of the Los Angeles Times.

JODY JACOBS   *1/13/78   L.A. Times*

Poof Goes the Punch in Pasadena

Dr. and Mrs. Robert F. Christy are plotting a night of magic Sunday in their Pasadena home for the trustees of Caltech.

After finishing some heavy research, Mrs. Christy has decided to work the dinner and the evening around a magic flaming brew called Feuerzangnbowle, which translates freely into fire-handled punch. That ought to warm up the party plenty. Since the punch is part of old German folklore, it naturally comes with a ceremony. One part of it involves a 15-minute show of lights as drops of sugar fall into the punch bowl from a burning, brandy-saturated sugar cone held high above it.

Each guest is to receive a sealed walnut shell with an individual magic message. And to go along with all this hocus-pocus the hostess will decorate the house with bubbling witches' pots and Caltech's student group, the Apollos Singers, and soloist Greta Davidson will sing numbers like "That Old Black Magic" and selections from Gilbert and Sullivan's "The Sorcerer."

We haven't quite given the whole game away, so there'll be plenty of surprises to keep trustees like Robert O. Anderson, R. Stanton Avery, Dr. Lee A. Du Bridge and Harry Volk and their wives on their toes.

Others looking forward to the magical evening are Mary L. Scranton who's coming in from Pennsylvania, Dr. and Mrs. Arnold O. Beckman, the Fred L. Hartleys, Mr. and Mrs. Franklin G. Larkin Jr., the Thornton Bradshaws, Mr. and Mrs. John A. Stephens, Mrs. Norman Chandler and Mr. and Mrs. F. Daniel Frost. Herbert L. Hahn has also accepted and so have Dr. and Mrs. Lawrence A. Williams, the Deane F. Johnsons, Mr. and Mrs. J. Paul Austin, Mr. and Mrs. Chauncey J. Medberry III, Augustine B. Kinzel and even more.

the dinners. I always included some appropriate faculty members among the dinner guests.

POOF -- AND WE HAD MUSIC!!

Soloist Greta Davidson sang selections from Gilbert and Sullivan's "The Sorcerer." She was accompanied by her husband, John, on the guitar.

She changed the words to "How to Handle a Woman" to fit the occasion:

How to handle the trustees,
There's a way, said a wise old man.
A way known by every President since
  the whole rigmarole began.
"Do I flatter them?" I begged him
  answer.
"Do I threaten or cajole or plead?
Do I brood or play the gay romancer?"
Said he, smiling, "No, indeed."
How to handle trustees,
Mark me well, I will tell you sir.
Now, "The way to handle trustees
  is to love them,
Simply love them, merely love them,
  love them, love them!

* * * * * * *

The Apollo Singers, Caltech students from the Glee Club, sang a beautiful selection of songs, among them "That Old Black Magic."

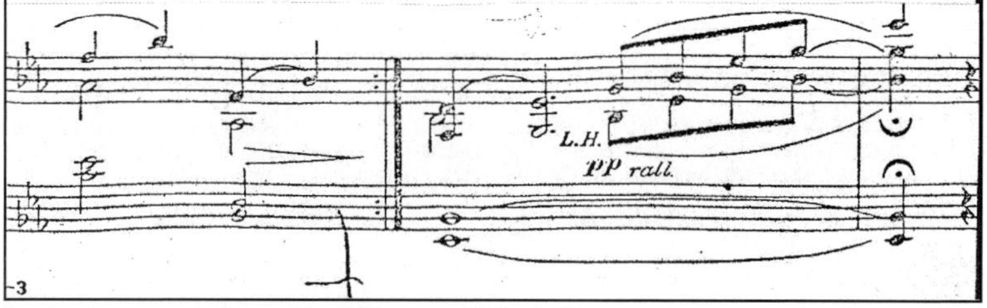

Lyrics to a song sung at the "Dinner of Magic"

The Trustees loved these dinners, and I tended to have a full house whenever I sent out invitations. Many of the Trustees were accustomed to having anything that money could buy, but they appreciated the personal attention that I gave to their visits. I always felt that the Trustees' most precious possession was their time. I was going to provide the most for the time they gave to Caltech.

Stan Avery really appreciated my entertaining and my other work on behalf of Caltech, and earnestly expressed this at every opportunity. So did Stanley Rawn, who once wrote, "Juliana is in a class of her own." Stan Avery eventually gave Caltech the

Robert Christy makes a presentation to the eminent physicist Max Delbrück (Caltech photo)

Commencement speaker Max Delbrück (standing at center) conversing with Trustee Stan Avery (Caltech photo)

The luncheon at the Christy house honoring the physicist Max Delbrück, who was the commencement speaker that year. Above left: Max Delbrück and Dietland Liepmann. Above right: Trustee Shirley Hufstedler and Max Delbrück. (Caltech photos)

The luncheon at the Christy house honoring Max Delbrück, showing guests at the tables that were set up in the garden behind the house (Caltech photo)

funds for a building, a large and striking student house.

Among the other duties that I enthusiastically took on was serving as a host to distinguished scientists who visited Caltech. I hosted monthly seminars for the Caltech Associates and the neighborhood. I also accepted invitations to many functions where I represented Caltech.

The Keck Telescope

I encouraged interest in Caltech among the Trustees and Associates in a number of ways. My professional background was in astronomy and astrophysics. I shared my fascination with astronomical objects by designing a special Christmas card with an image

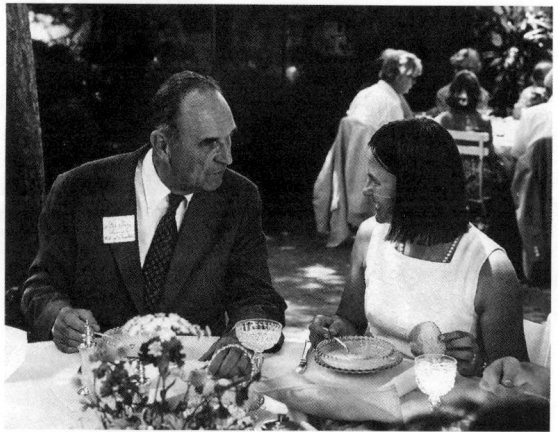

Trustee Stan Avery and Juliana Christy, at her luncheon for Max Delbrück (Caltech photo)

of the beautiful Pleiades star cluster. This was the official Caltech card sent to thousands of people for Christmas 1977.

I had been accustomed at the University of Toronto's David Dunlop Observatory to give three talks each week to the public, in order to encourage interest in astronomy and to demonstrate what could be seen in the sky with the 74-inch telescope. This skill that I had developed at the University of Toronto proved extremely useful at Caltech. At one weekend-long Trustee meeting in Palm Springs, I guided 80 of the Trustees and their spouses in a walk out into the desert late one evening. It was a long

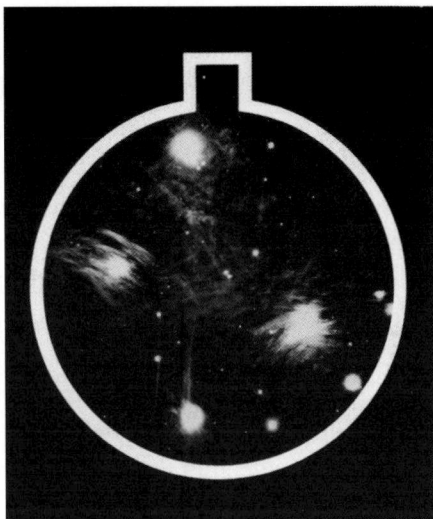

The Christmas card showing an image of the Pleiades

column of people following me, after dark when the stars were clearly visible. I had invited several Caltech professors to join us on this night walk, to help guide the long column of Trustees and their spouses through the desert, and to be readily available to answer questions from anyone who showed interest. I remember Maarten Schmidt and Robbie Vogt as being two of these Caltech professors. I showed the Trustees what one could see of the sky with the naked eye. Then I brought out about half a dozen pairs of binoculars and showed the Trustees how much more could be seen with these, including planets, double stars, clusters of stars, and the Andromeda galaxy. Andromeda is the nearest spiral galaxy, a neighbor to our own Milky Way galaxy. It looks very similar to what our Milky Way galaxy would look like if we could view it from the outside.

I tried to inspire the Trustees by talking about some of the other wonders that were out there. Beforehand, I had prepared for this night's presentation by making (by hand) 80 booklets containing beautiful color photographs of what had already been seen by Caltech's present powerful telescopes. I explained to them what else there might be out there to be seen, if only one had an even more powerful telescope to see it.

It was Robbie Vogt who continued the efforts to interest the Trustees in astronomy, in the hopes of getting a new and larger telescope for Caltech. Trustee Bill Keck was one of the Trustees who had been on the above desert walk; he had been inspired by my talks about the beauty and mystery of space, and of the great need for high caliber telescopes to reach into the depths of the universe. Following his death, his brother provided half of the funding for a major telescope — $70 million — in honor of Bill Keck. This was to be called the William M. Keck Telescope. This 10-meter telescope was the world's biggest telescope at the time that it began operation, in 1993.

This put Caltech in the lead again in terms of telescope size and the resulting ability to observe the faintest and most distant objects in the universe. The time it takes light to travel from these distant objects means that the light we observe today left these objects in the

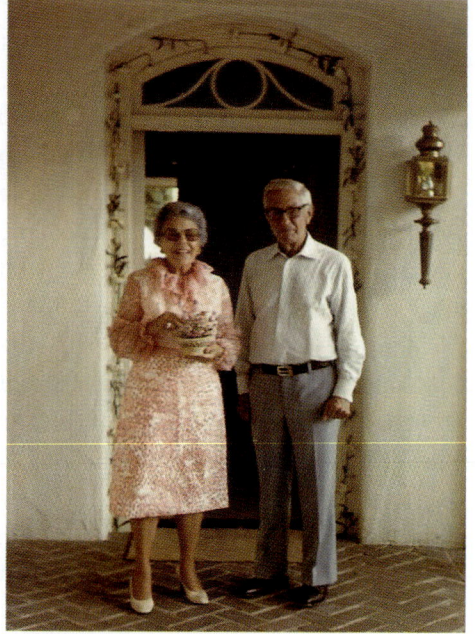

Arrola and Lee DuBridge at the entrance of the Christy home

far past, not long after our universe began. Thus one can see not only into far distant space, but also into the far distant past.

Caltech had been in a similar leading position in 1949, when the 200-inch (5-meter) Palomar Telescope began operation.

Caltech's Oriental Rugs and Other Art Treasures

I looked into the preservation of student houses, and tried to protect treasures such as priceless tapestries and Oriental rugs at Caltech. For example, Caltech's Athenaeum had the second-largest oriental rug in the world. I located other oriental rugs in the student dormitories, totally unprotected and un-cared-for. The Caltech President's house had several such rugs, and some were also to be found in other commonly-used buildings such as Dabney Hall. In an attempt to organize cleaning and repair, and to prevent thefts, I initiated the first appraisal and tabulation of Caltech's tapestries and Oriental rugs. I did not have time to complete this task before Robert decided to relinquish the President's job, but a preliminary list found 27 Oriental rugs. I also initiated an attempt to create a committee to recommend ways of protecting, enhancing, and keeping an inventory of Caltech's vast collection of fine arts. These included not just the Oriental rugs but also such items as paintings, antique furniture, vases, and urns.

Cutting Administrative Expenses, and a New Source of Funds

Robert and I tried to be an example for saving money at Caltech in various ways, so that it would be available for students and faculty. Wherever I could, I tried to save funds that the President and his wife had available to them. For example, I never spent money for professionally made flower arrangements for dinners, luncheons, commencement, etc., but rather made the flower arrangements myself with the help of my secretary, Barbara Clark. We gathered flowers early in the morning from the Caltech campus, and if there were not enough in bloom, we supplemented these by buying selected flowers from flower shops. I never flew first class at Caltech's expense, and in fact did not travel much at all — though that was partly because I had just given birth to our second daughter.

From Robert's dinner conversations at home, I was very much aware of how badly Caltech was in the hole financially at that time. I had an idea to help. I knew of an income-generating building for sale close to Caltech, at 1111 Blanche Street, owned by the movie star Dean Martin. It had forty-four small but luxurious two-bedroom two-bath apartments. There were several levels of management taking care of the building, but it was terribly mismanaged, with about 25% vacancy (ten apartments vacant). At $190 per month, the rents in the building were about half of market value at the time.

Due to the vacancies and the low rents, this building could be purchased at about half its market value. Since it was close to Caltech, it would be convenient housing for Caltech graduate students and faculty. I introduced this investment opportunity to

At a Caltech commencement: Juliana (left center) sitting between Jean Bacher (left) and Arrola DuBridge (right center)

David Morrisroe, who was
Caltech's vice president for
financial affairs and treasurer.
It was an investment gold
mine — inflation-safe, since
rents go up with inflation —
and I persuaded him to buy
it for Caltech, for about
$800,000. With rents today
at $1600 per month (still
somewhat below market
value), the building presently
can provide a net income of

Arrola DuBridge, Juliana Christy, and Ned Munger
(Caltech photo)

over half a million dollars a year for Caltech. These funds are precious because
they are not earmarked for some specific purpose, but are available wherever
there is the greatest need. David Morrisroe used to say that this was one of
Caltech's best investments.

When the search committee was looking for a new President and First Lady
for Caltech, I wrote up a guideline of what the First Lady's job really entailed,
and what the President's secretarial staff could not fulfill. The search committee
was very grateful, as they had not had any clue as to what was involved. I also
suggested to them in writing that if the wife of the new President had had a
paying position prior to her husband accepting the Presidency of Caltech, it
would be in Caltech's best interest to pay such a woman at least half-time. Until
then, they had always taken it for granted that the wife would do everything for
free, and would have no income in her own person for what she was doing. My
attempts in that direction were not in vain; eventually, the incoming First Ladies
of Caltech were treated the way I had suggested. The First Lady of Caltech is not
just a wife: she has an important and influential position of her own to fulfill.
Her efforts as a gracious hostess are vital to Caltech's fundraising activities, and
to engendering goodwill towards Caltech in Washington and elsewhere.

Offers of Presidencies from Other Universities

During the period when Robert was Interim President of Caltech, he was offered
the Presidencies of several other universities, including the University of British
Columbia in Vancouver, Canada. Vancouver was my favorite city to live in, so I

would have pushed for this if I had known of it. However, unlike with his first wife, Robert never mentioned this possibility to me since he had no desire to leave Caltech.

With his characteristic modesty, Robert never mentioned these offers of university Presidencies to anyone else in the administration, such as the Trustees — most people would have at least used such offers to bargain for something better at their present institution. Despite his reticence to mention these offers, the Trustees noticed his abilities and asked him whether he would agree to continue as Caltech's President for a full term in office, which in those days was eight years. He declined. He then did everything he could to assist in finding a new President for Caltech.

Juliana's Choice

After Robert declined the Trustees' offer of Caltech's Presidency, they asked me whether I would be willing to continue as First Lady of Caltech, since they had one capable candidate who was divorced and had no lady to carry out those of the First Lady's tasks that a secretary could not do. They reminded me about Jefferson, the third President of the U.S., for whom Dolley Madison (wife

of James Madison, Secretary of State) had sometimes acted as First Lady. I, however, felt that I could not carry out the task of First Lady without the private discussions about events that one tends to carry out in the privacy of one's home when one is married. Even though I would have liked to continue in the job, I felt that without my husband's involvement I should not do it.

The Actor in the Acting President

In 1977, while Robert was Interim President of Caltech, he was also an actor in a play — "Fiorello!" — that was presented at Caltech. In it he played a radio announcer. Robert's colleague

Robert Christy as the radio announcer in the play "Fiorello!"

Dick Feynman played the character Frankie Scarpini, as well as one of the "Italian Street People" in the chorus. Dave Morrisroe, Caltech's Vice-President for Financial Affairs and Treasurer, played the First Heckler.

Coincidentally, there also existed a professional actor, Jimmy Stewart (1908-1997), whose appearance and personality were strikingly similar to Robert Christy's.

On the trip to China: Fifth from left is Arnold Beckman, then Stan Avery, Seymour Benzer, Robbie Vogt, Robert Christy, Barclay Kamb, and unknowns (Caltech photo)

A Trip to Red China

It had been President Nixon's 1972 visit to Red China that had first established diplomatic relations; prior to that, Americans were not permitted to travel to China. Six years later, in 1978, Robert Christy travelled with a Caltech delegation consisting of Trustees, Caltech administrators, and a few professors to Red China, to establish contacts in science. In a postcard to me and our little daughters, Robert wrote:

We leave Peking tomorrow for Sidn. I purchased a small (2'x4') rug for your mother (OMA) and mailed it to her. It is very attractive and I intend it as a gift.

There are millions of people on foot & on bicycles going somewhere all the time with a few cars honking. It is a country of contrasts. People living in peasant like conditions just behind a major modern hotel is typical. A lot of construction going on, much with hand labour. We have been very busy but are having a fantastic experience.

Love to all,
Bob.

On the trip to China: from the right: Robbie Vogt, unknown, Arnold Beckman, Robert Christy, Stan Avery, Seymour Benzer, unknowns; at left: Barclay Kamb (Caltech photo)

CALIFORNIA INSTITUTE OF TECHNOLOGY

PASADENA, CALIFORNIA 91109

OFFICE OF THE CHAIRMAN
OF THE BOARD OF TRUSTEES

October 12, 1978

RECEIVED

OCT 16 1978

OFFICE OF THE
PROVOST

Dr. Robert F. Christy
California Institute of Technology
Pasadena, California 91107

Dear Bob:

I am enclosing a copy of the photograph with Vice Premier Wang-Chen which Madame Wang gave me to pass along to you as I was leaving Peking.

We don't look as wet and bedragled as we actually were from passing through the rain storm on the way to the Hall of the People. I certainly enjoyed the experience and Seymour has reason to be especially grateful for this photograph having been published in Renmin Piao.

I thought the trip was a great success standing by itself, but was a particularly good way of getting all of us acquainted in a way that is difficult to achieve short of going down the Colorado River or all the way to China.

I was sorry to miss Chou Pei-yuan's and Dr. Weii's visit this week but Murph said he would make my apologies if needed. It will be interesting to see what develops from here on out.

It was great to be with you.

Sincerely,

Stan

Enclosure

A letter about the China trip from Stan Avery, Chairman of the Caltech Board of Trustees

Photograph with Chinese vice premier Wang-Chen while on trip to China. <u>Front row:</u> 2nd from left is Seymour Benzer, 4th is Robert Christy, 6th is Murph Goldberger, 7th is Stan Avery, 8th is Arnold Beckman, 9th is Robbie Vogt. <u>Back row:</u> 3rd from left is Bruce Murray, 4th is Barclay Kamb.

The Goldberger Presidency

It was a difficult time to be searching for a new President for Caltech. Jimmy Carter, the first Democratic President in Washington in eight years, had attracted many of the capable professors (who on average tend to be Democrats) to his government positions to help run the country. It was very hard to find an appropriate person willing to run a prestigious but small institution such as Caltech. Finally, after having been turned down many times by various candidates, Caltech's Presidential Search Committee came up with Marvin ("Murph") Goldberger. He had never had any major administrative experience; he had only been leader of a small group of physicists at Princeton. However, he was the only candidate who enthusiastically wanted the job. There was a crucial secret meeting in which the Trustees voiced their concerns. Robert Christy was called in at the end, and asked for his opinion of Goldberger. He said that he knew Goldberger to be a good physicist. Partly due to that statement and Robert's credibility at Caltech, Goldberger was offered the Presidency of Caltech early in 1978. Murph Goldberger enthusiastically came to Caltech to look everything over.

The President always has the privilege of selecting his own Provost, so as to have a second-in-command with whom he can work comfortably. In a private meeting with Robert immediately after Goldberger was offered the Presidency of Caltech, Goldberger informed Robert that he did not wish to work with him as Provost. Robert was to continue working as Provost until someone else could be found for the position.

Provost Robert Christy with new Caltech President Marvin L. Goldberger, at a reception in the Goldbergers' honor (Caltech photo)

At this point, after eight years, Robert had become comfortable with being Provost. However, as requested by Goldberger, Robert continued in that post for only another two years, until a new candidate could be found. After two years of working with Goldberger, Robert Christy had also concluded that the two of them were incompatible.

John D. ("Jack") Roberts became the third Provost of Caltech in 1980, following Robert's Provostship. However, Jack only remained as Provost for

The first four Provosts of Caltech: 4th Provost Robbie Vogt, 1st Provost Bob Bacher, 3rd Provost Jack Roberts, and 2nd Provost Robert Christy (Caltech photo)

three years, quitting the position in 1983. Jack was one of Robert's best friends at Caltech — they regularly played tennis with each other. For a number of years, Robert used to go skiing with Jack, his wife Edith, and their four children at Aspen and at Mammoth. Jack had been the best man at our wedding.

After Jack Roberts resigned from his position as Provost, Robert Christy talked to Robbie Vogt, who was the Chairman of the Division of Physics, Mathematics, and Astronomy at Caltech. Robbie Vogt enjoyed administration, and was outstanding at it when he put his mind to it. He had worked very hard as the Division Chairman, and was one of the best that Caltech had ever had, similar to Bob Bacher. Robert encouraged Robbie Vogt to serve as Provost, believing that he would be very good at the job — and so he was. Among other accomplishments he was instrumental in making the Keck Telescope a reality. This telescope was the largest in the world at the time, aiding Caltech and the telescope's co-owner UCLA to be in the forefront of astronomical discovery.

It was in 1980 that Robert left his administrative position as Provost and returned to teaching and research. I had always felt that it was a tragedy that a brilliant scientist like Robert Christy should leave his research, where he excelled. Dick Feynman had the same opinion. Shortly after Robert took on the Academic Vice-Presidency of Caltech, at a chance meeting on Caltech's Olive Walk, Feynman shook his head and said to Robert, "Christy, how could you ever leave physics and research and go into ad…min…is…tra…tion?" It was good for Caltech, but a sacrifice on Robert's part: another example of his generosity.

The Trustees' Reaction to Robert's Departure from Administration

After Robert's departure from administration, several of Caltech's Trustees offered to use their influence to obtain offers of major administrative jobs for Robert elsewhere, in Washington, DC and elsewhere in the U.S. Robert declined, since he wished to remain at Caltech. He became a professor of theoretical physics again, teaching courses and looking around for new research topics.

On the following pages are letters, most of them written in 1978 (at the end of Robert Christy's term as Interim President of Caltech), from Trustees Arnold O. Beckman, Thomas J. Watson, Jr., Stanley R. Rawn, Jr., Augustus B. Kinzel, Mary Scranton, Bill Fitzhugh, and from the previous President of Caltech, Harold Brown, who was then the U.S. Secretary of Defense.

ARNOLD O. BECKMAN
107 SHORECLIFF ROAD
CORONA DEL MAR, CALIFORNIA

RECEIVED

MAR 1 3 1978

OFFICE OF THE
PRESIDENT

March 7, 1978

Dear Bob:

Now that the uncertainty concerning the Caltech presidency
has been resolved, it seems appropriate to look back over the
months that have elapsed since Harold Brown's departure, to
see how the Institute has fared.

My personal view, which is shared by other trustees, is that
things have gone well, a result in large part of your leadership.
You have done an excellent job as interim acting president. The
trustees, faculty and students owe you a large debt of gratitude for the way in which you handled problems that arose and
kept operations going smoothly. All in addition to discharging
your regular duties as provost!

I warmly commend and thank you and Juliana. I hope that the
experience provided some gratifying rewards to you both.

Sincerely,

Arnold

Note Robert Christy's handwritten draft of his reply on this letter from Arnold Beckman

Thomas J. Watson, Jr.
Old Orchard Road, Armonk, New York 10504

March 14, 1978

Dear Bob,

Now that your job as Acting President of
Caltech has come to an end, I just wanted to
write you this note and let you know that I
think you and Mrs. Christy have done a great
job for the last 14 months under extremely
difficult circumstances. Your balanced
approach to the various problems which presented
themselves during this critical period has
given me great admiration for your diplomacy
and ability.

Best regards,

Sincerely yours,

Dr. Robert F. Christy
California Institute of Technology
Pasadena, California 91109

STANLEY R. RAWN, JR.
645 MADISON AVENUE
NEW YORK, N. Y. 10022

RECEIVED

MAR 15 1978

OFFICE OF THE
PRESIDENT

March 13, 1978

Dr. Robert Christy
California Institute of Technology
Pasadena, California   91109

Dear Bob:

      I know you will be receiving a barrage
of richly deserved accolades for your extraordi-
nary performance during the past year and a quarter.
I want to add my personal congratulations and thanks.
Caltech is most fortunate to have you - and to that
I want to add Juliana. As Caltech's hostess, she
is in a class by herself. All those associated in
any way with Caltech owe you a great debt of gratitude.

      Well done!

               Warmest regards,

               Stanley R. Rawn, Jr.

SRRjr/es

AUGUSTUS B. KINZEL
1738 CASTELLANA ROAD
LA JOLLA, CALIFORNIA 92037

11 March 1978

Mr. Robert F. Christy
1230 Arden Road
Pasadena
California   91106

Dear Bob:

Now that we have elected a President, I want to tell you how much I have admired your handling of the interim post. I know it required an unusual degree of wisdom, tact, and devotion. In addition, your offer to continue for a reasonable period will help the new incumbent. It is most gracious and generous.

Of course I hope we can keep you at Caltech, but if you should decide otherwise, be sure to feel free to call on me as a reference or in any other way in which I could be helpful. I know that many other Trustees have greater influence in many quarters, but I do happen to be close to the Washington scientific community.

Again, my admiration and thanks, and all of the best to you and your charming wife.

Sincerely,

Augustus B. Kinzel

Achieving the Rare

March 17, 1978

Dr. Robert Christy
California Institute of Technology
Pasadena, California 91125

Dear Bob:

Thank you so very much for all that you have done
for Caltech this past year.

You have certainly earned the admiration and respect
of all the Trustees for a very difficult task and I hope
that you realize how much the Trustees appreciate what
you have done at a difficult time.

Although the Board has tried to express its appreciation, I would like to add my personal thanks for all
that you have done and are doing.

Bill and I have just returned from our vacation in
Arizona and find that we are now getting six more inches
of snow. Perhaps I will be out in Pasadena more frequently
than I planned - at least if this weather keeps up I am
sure I will be!

With all best wishes always.

                    Most sincerely,

                    Mary

                    Mrs. William W. Scranton

Box 1452
Rancho Santa Fe, California 92067
March 6, 1978

Dr. Robert Christy
Acting President

Dear Bob,

I just wanted to put in writing what I said orally after this morning's meeting — You've done a great job under difficult conditions, and deserve — and will get — the hearty thanks of all people having an interest in Caltech's prosperity. You never made a false move!

Van and I wish you both every happiness and a long future at Caltech.

Sincerely,

Bill Fitzhugh

"Dear Bill: Thank you for your warm note. I have only realized only after the fact the peculiar problems of being acting President and I particularly appreciate the fact that you and the Trustees are sensitive to this situation.

Sincerely,
Bob.

Again, Robert has written a draft of his reply at the bottom of this letter from Bill Fitzhugh.

```
CALTECHGAF PSD
PP 910-588-3255   V 510-765-6678  NR038A/27FEB80/DOD
P 270124Z FEB 80
FM SECDEF WASHINGTON DC
TO CALIFORNIA INSTITUTE OF TECHNOLOGY
   PASADENA CA 91125
   ATTN MR HARDY C MARTEL
BT
UNCLAS   1,45
WHILE UNFORTUNATELY I CANNOT PERSONALLY BE PRESENT AT THIS SPECIAL
DINNER IN HONOR OF BOB AND JULIANA CHRISTIE, MY THOUGHTS ARE WITH YOU
THIS EVENING.  BOB'S MANY YEARS AT CALTECH HAVE EARNED HIM THE HIGH-
EST REGARD OF ALL THE INSTITUTE COMMUNITY.  I, AND ALL OF YOU AS
WELL, HAVE PARTICULAR REASON TO BE GRATEFUL FOR THE YEARS DURING
WHICH I RELIED SO MUCH ON HIS OUTSTANDING PERFORMANME AS PROVOST.
HIS CALM AND EVENHANDED APPROACH, THOUGHTFUL CONCERN FOR INDIVIDUALR,
AND HIS STEADFAST DEVOTION TO THE INSTITUTE HAVE SET A STANDARD FOR
ALL TO EMULATE.  HE WILL, I AM SURE, CONTINUE TO LEAVE HIS MARK AS HE
RRUERSSOROIL LIFE.
CSHESETANBOB SNNDJOURANERFOBESTAT
THEY HAVE DONE ANL FOR WHAT THEY HAVE YET TO DO.



PAGE  2 RUEKJCS 2205 UNCLAS
HARO D BROWN
BT
#2205




NNNN
REPEAT OF GARBLED LNS ABOVE

RETURNS TO PROFESSORIAL LIVE.
COLENE AND I SEND OUR VERY BEST WISHES TO BOB AND JULIANA FOR WHAT
NNNN
CALTECHGAF PSD
```

A telegram from Caltech's former President Harold Brown and his wife Colene

To

Robert F. Christy

from

The Faculty of the
California Institute of Technology

in appreciation for

his dedication and patience

in shepherding us with dignity

through a long, and in some ways

trying, interregnum

Presented at the faculty dinner
May 24, 1978

James J. Morgan
James J. Morgan
Chairman of the Faculty

Chapter 15

Later Years — Teaching and Research

Robert was only 64 when he left his administrative work and returned to teaching and research. In his 2006 interview with Sara Lippincott, Robert recalled:

> I went back to teaching for a while. But I didn't find it so easy again, because after being out of it for eight or ten years, you begin to lose touch. So I wasn't as satisfied with the teaching as I should have been. I don't think I did it as well. And I was getting close to retirement, too. So I gradually retired [and became emeritus].
>
> Now, during that period, I did try to learn a new field. Because I have always enjoyed exploring a new field to see whether I could find something where training in physics could make a contribution. And a field that I found fascinating then — and I'm still fascinated by it — is the question of what caused the oscillations in climate on the Earth that led to the Ice Age. And as you know, 25,000 years ago, there was an awful lot of ice in the Northern US and Canada. And this was not understood. I did a lot of reading in this field — climate and so forth. I did not find any real key that I felt would unlock it. Although there were theories being propounded, I felt the ones I saw were really not fully adequate. And yet I didn't find anything better.

Satellite data concerning the ocean currents had become available at that time, especially via JPL, which is closely associated with Caltech. Robert hoped to use this data to gain a better understanding of how the oceans influence the climate. Robert said:

> My approach is usually to immerse myself in the facts, so that I can then see what the constraints are. I read from time to time about the work that people have done on the circulation in the oceans, which has very profound influences on climate and has barely been worked out yet. Very complex. I don't know yet what the answer is, but I still find it an interesting field. Climate is kind of a persistent weather pattern — persistent for hundreds of years; whereas weather is just what goes on from day to day, or month to month. People have learned a lot about weather, and they have found that it's technically unpredictable. You can predict for a few days, but you cannot basically predict for much more than a week. Because no matter how big a computer, you won't be able to make the predictions

go much longer. But climate is different. There are certain patterns of weather that are persistent for years that have nothing to do with your ability to predict from day to day. It's a superimposed pattern.

In the end, Robert decided that he was not making sufficient progress on climate prediction and gave up this line of research. Fortuitously, in 1981 he was invited to be the Chairman of a major project that he was uniquely qualified to work at, and that would give him the satisfaction of being able to contribute. This was the dosimetry project described in Chapter 16.

Robert's Titled Professorship

In 1983, the Trustees of Caltech wished to recognize Robert's many contributions, and bestowed a titled professorship on him, "Institute Professor of Theoretical Physics." On the next page is a letter from Lee DuBridge, who had known Robert since the 1940's. He had been Caltech's President for 20 years, and was now a Trustee.

Robert's Retirement from Teaching

In 1986, when Robert turned 70, he retired from teaching (although he continued his scientific work, mostly outside Caltech). On the following pages are letters written to Robert at this time, from Shirley M. Hufstedler, Trustee of Caltech; R. Stanton Avery, Trustee of Caltech; and Bob Sharp, a Caltech colleague.

Lee A. DuBridge
1730 Homet Road
Pasadena, California 91106

June 28 1983

Dear Bob:

Arrola and I were
delighted to read the
announcement of your appointment
as Institute Professor. This
is an honor long overdue —
and certainly richly deserved.
Your *many many* friends
will be greatly pleased.
Congratulations!

Lee & Arrola

Hufstedler
Miller
Carlson &
Beardsley

August 14, 1986

Dr. Juliana and Dr. Robert F. Christy
W.K. Kellogg Radiation Laboratory
California Institute of Technology
Pasadena, CA 91125

Dear Juliana and Bob:

    I have only just emerged from a crunching load of
work that has kept my head down since last Christmas.
Therefore, at long last, I can express, albeit very too
briefly, my admiration of Bob's career, his contributions to
CalTech, and my hopes that many rewards will come to both of
you in the years following retirement.

    One of the pleasures of my service on the Board has
been the opportunity that it has given me to become acquainted
with both of you and, from time to time, to work with both of
you on specific projects for CalTech. I hope before long that
I will have a little free time which will permit me to take
both of you to lunch where we can catch up with all your doing
and the activities of the girls.

                                  Sincerely,

                                  SHIRLEY M. HUFSTEDLER

SMH:lhn

Avery International

R. Stanton Avery
Founder Chairman

May 16, 1986

150 N. Orange Grove Boulevard
Pasadena, California 91103
Phone 818/304-2151

Mrs. Robert F. Christy
1230 Arden Road
Pasadena, CA 91106

Dear Juliana:

It was very thoughtful of you, Juliana, to send me
a copy of Bob's story. I was really not at all
acquainted with the details of Bob's background and
was very pleased to learn more than I had known before.

This is a wonderful story of a wonderful life.
Bob still does not look old enough to have accomplished
all those things and lived all those years. I know
that he will go on and on and on with great pride in
his family and you, Juliana.

Sincerely,

Stan Avery

CALIFORNIA INSTITUTE OF TECHNOLOGY

DIVISION OF GEOLOGICAL AND PLANETARY SCIENCES 170-25

May 14, 1986

Dear Juliana:

"I appreciate ever so much your warm note of May 13th and the enclosed copy of your notes to Bill Fowler.

Everything you say about your Bob is absolutely true, even more so. Yes, he is like one of those towering gorgeous trees of his native land. Quiet, but oh so strong, straight and upright. His 24 carat integrity, impeccable judgement, razor-sharp mind, and devotion to Caltech are beyond compare.

Working with him was always a pleasure because of those characteristics

(PTO)

PASADENA, CALIFORNIA 91125 TELEPHONE (818) 356-6811

and because he was always a pillar
of strength and good common sense
in the midst of confusion.

. I much regret missing his
retirement dinner, but I am that
night off with a group of students
on one of our field trips. My
disappointment is partly compensated
by the fact we will be (part of the
time) in an area dear to the both
of you — The Bridgeport Basin. You
both will be much on my mind as
we pass the Hunewill Ranch.

Robert is an incredible man

Faithfully yours
Bob Sharp

Robert at his "official retirement" dinner with daughters Alexa and Ilia, and Vera Nicholas (wife of his son Ted, at right behind the flower arrangement)

Robert's "Retirement" Dinner: Still Amazingly Young at Age 70

Most faculty retire at age 65 or 70. Caltech hosted a major retirement dinner for Robert when he was 70 years old. Two photographs show Robert at this event. It is noteworthy that he had no gray hair and looks amazingly young. He appears younger than in photos from fifteen years earlier, when he was in his mid-50's and was going through the stress of the failed marriage with his first wife. A happy marriage and an active, fulfilling life is crucial in helping to keep one young.

Robert Christy at age 70 at his "official retirement" dinner. Back: Robert to the right of his son Ted. Middle: Juliana's mother Lilly Stelter, Ted's wife Vera Nicholas, and Juliana. Front: Robert's daughters Ilia and Alexa with Prof. Willy Fowler between them.

Inertial-Confinement Fusion

In the 1990's, when Robert was already in his 70's, he worked for a number of years on a Department of Energy committee to review "Programs in Inertial Fusion." In his 2006 interview with Sara Lippincott, Robert recalled:

> For quite a number of years now, I have been on one committee after another that has been pursuing a review of what's called inertial-confinement fusion. You usually hear about magnetic fusion. Fusion is where you somehow or other bring hydrogen atoms together to make helium. In magnetic fusion, you use magnetic fields, very intense magnetic fields, in very difficult configurations called tokomaks, to confine the hydrogen isotopes.
>
> What I've been looking at is inertial-confinement fusion, where you try to make a very tiny hydrogen bomb — a hydrogen bomb less than a centimeter across. You don't use magnetic fields to hold the stuff together — it's held together only momentarily. It's driven together by an outside force — usually lasers. Very powerful lasers shine on this little pellet of material and drive it together very hard. And in the instant when it's most compressed it may — if you've done it right — fuse and detonate. It's hard to make them [the explosions] very powerful. And, of course, it's hard to make them very weak — if you make them at all.
>
> It's thought of as a reactor-type device. The energy would then be contained in something probably ten or fifteen feet across. Then cooling material would take the energy out, and you would have an energy source. But it's very difficult to do. So there have been committees appointed, supported by the Department of Energy, appointed through the National Academy or the Department of Energy, reviewing our progress. Are we pursuing this in a sensible way? Should we continue? This work is being done at Los Alamos, Livermore, some work at Sandia, some work at Rochester, some work at the Naval Research Laboratory in Washington. And we've been going to these various places and trying to advise the Department of Energy as to where to put its effort in this direction, and what the prospects of success are. It is an alternate approach to magnetic fusion. Neither one is easy. Magnetic fusion is exceedingly difficult — both to make things fuse, and then to get the energy out — because of this enormous strong magnet they have to use. And the whole problem with inertial confinement is to make more power when you implode the pellet than you use up with your lasers. I think you could do it fairly easily, if you were willing to put a few tens of millions of joules into laser energy and not worry about whether you got it back.
>
> If you start with a very small amount of material — deuterium — in the way of milligrams or something, you can easily calculate that if you do succeed in making it fuse, it doesn't make that much energy. [So it is pretty safe.]
>
> Supposedly fusion energy is clean energy. That's what they say. But it's a relative question. Fusion energy, for the same amount of energy, does make less

Presented to

Dr. Robert F. Christy

This certificate is awarded in appreciation of your contributions to the success of the ICF Program. The accomplishments, based on the ICF program staff's dedication and talent, have led to DOE support of the National Ignition Facility and to a bright future for ICF. The Inertial Confinement Fusion Advisory Committee (ICFAC) helped provide a focus, a sense of urgency, and a forum for discussion and debate of contrasting points of view.

November 1995

radioactivity than fission does. But on the other hand, it still makes tens of thousands of times more radioactivity than we can stand. So the radioactivity made is one of the major problems with fusion energy as well as with fission energy. A difference in the radioactivity is that fission makes radioactivity of short life, intermediate life, and very long life. With fusion, it makes short-life radioactivity and some intermediate ones, but not as much in the way of exceedingly long-life radioactivity. The same problems will exist, but not as severely. But since people find [the radioactive waste problem] almost insoluble with fission reactors, that doesn't mean that a little bit less severity is going to solve things. It's going to be very difficult to deal with radioactivity. And it's also very difficult to get the energy out. It's very difficult to engineer either type of fusion [magnetic or inertial confinement] to make it work.

Every time we meet and discuss the possibilities of success, it's always at least twenty years down the road. And it's been going on that way for the last forty years — fusion has always been twenty years down the road. And it still is at least twenty years down the road, if you can make it work at all. But technically, it's most challenging. The lasers are fantastic. And all the technology is fantastic. The calculations that people do are fantastic. It's very interesting to find out what people are doing.

The technical problems are major. It's far more difficult than fission. Fission was like falling off a log; you could hardly *avoid* making a fission reactor. Fusion is very unnatural. In the center of the sun, it works naturally. But it does not work naturally on Earth. So it is exceedingly difficult, technically, to make it work. Whereas, it's very difficult to avoid having fission work. So there's a very great difference in the technical difficulty: very high temperatures that cannot be contained permanently, except by strange methods such as magnetic fields and so forth. It's not something that is at all straightforward.

A Consultant for the History He Had Witnessed

For many years, Robert had been an occasional consultant for companies who valued his expertise in physics. Late in his life he was sought after as a different kind of consultant, as an expert on the history of the exciting events that he had been part of. As one example of this, Robert was among both the first and the last people interviewed in a frequently-shown episode of "Modern Marvels," produced for TV by the History Channel. I remember him being on a number of other TV programs of which we have no written records.

Robert participated in the production of the "Critical Mass" exhibition at the Museum of Fine Arts in Santa Fe, New Mexico. This exhibit brought together photography, video, and text to examine the forces that led to the creation of the first atomic bomb, and was on display at the museum from November 6, 1993 to February 13, 1994.

Robert also consulted for a 2003 production by the British Broadcasting Corporation titled "To Mars By A-Bomb," about the Orion spacecraft project on which

> This certificate expresses a heartfelt THANK YOU to
>
> *Dr. Robert Christy*
>
> for participating in the production of the installation
>
> *CRITICAL MASS*
>
> Meridel Rubenstein
> Artist
>
> Ellen Zweig
> Artist
>
> David Turner, Director
> Museum of Fine Arts

Certificate of appreciation from the Museum of Fine Arts in Santa Fe, New Mexico

he had worked briefly with Freeman Dyson (as described in Chapter 9).

In 2010 a play was written about Robert himself by Andrew Kerwin, a student at Fruitvale Junior High School in Bakersfield, CA. The project had been initiated by the school's Principal, John Hefner. This play, "The Manhattan

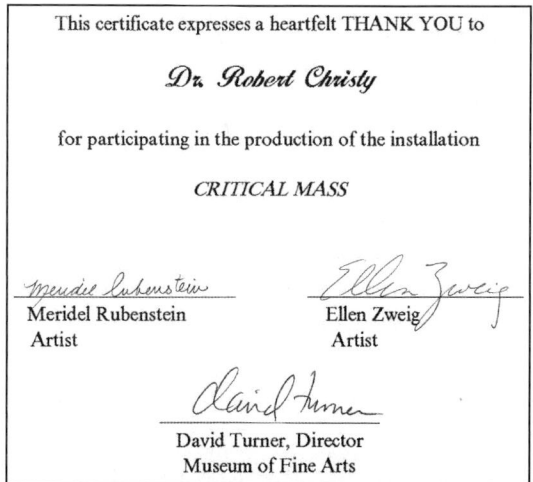

Project: An Innovation that Would Change the World," won a History Day competition in Bakersfield. It went on to win California's state-wide competition, and placed among the top ten in the national competition.

A number of playbills and letters from Robert's files provided a sample of some of the other projects in which he was involved, as summarized here:

Robert Christy, Andrew Kerwin (high school student and author of a play about Robert), Juliana Christy, and John Hefner (Principal of Fruitvale Junior High School, retired)

In a letter dated "Paris, 21st of November 1995," Producer Nicolas Blanc of Agat & Cie Films thanked Robert for his collaboration on the realization of Arthur Mac Craig's documentary "I am become death: They made the bomb."

In the fall of 1997 Robert appeared on the "Nuclear Power/Energy" episode of the Polygon Show. This was an informative talk radio program on the American Radio Network, co-hosted by Diane Moca and Marvin Feldman.

On December 10, 2001, Caltech held a public roundtable discussion called "The Copenhagen Interpretation: Exploring Science on Stage." Robert Christy was one of the five panelists who discussed the 1941 meeting in Copenhagen between Heisenberg and Bohr, and why Nazi Germany never built an atomic bomb. This was the meeting dramatized in Michael Freyn's play "Copenhagen." Heisenberg was the top scientist in the German atomic project, and was of course closely supervised in his visit to Bohr. Any unauthorized information could only be hinted at. It has been suggested that he may have been trying to reassure Bohr that he had no intention of constructing atomic weapons. However, Bohr interpreted his hints as a warning, and his report of this meeting added momentum to the Allied atomic project.

A letter dated May 7, 2007 thanked Robert for his "continuous support" to the NHK Japan Broadcasting Corporation. The letter mentioned that Robert had contributed to their 1998 English-language program "Hiroshima: The Fateful 10 Seconds." This program was to be broadcast again in 2007 on NHK World TV as part of the NHK Peace Archives Project, which had started in 2005 to commemorate the 60th anniversary of the end of the Second World War.

In the spring of 2008, Mark Ward of L.A. TheatreWorks asked Robert for an interview for an episode on the Manhattan Project as part of their science-themed radio series. This series was aired on KPCC and other public radio stations across the U.S.

Finally, in 2009, Robert was a consultant for the PBS film "The Trials of J. Robert Oppenheimer" by David Grubin.

Chapter 16

Radiation Dosimetry in Hiroshima and Nagasaki

In 1981, Robert Christy was asked to chair the working group of the US-Japan Joint Reassessment of Atomic Bomb Radiation Dosimetry in Hiroshima and Nagasaki. This project was a massive collaboration between the U.S. and Japan, involving scientists in relevant fields as well as physicians. They were trying to understand the long term effects of relatively low doses of radiation.

History and Purpose of the Project

In 1946 President Truman had approved a directive to the National Academy of Sciences to initiate a long-term study of the surviving populations in Hiroshima and Nagasaki. The Academy established the Atomic Bomb Casualty Commission (ABCC) in 1947 and the Japanese National Institute of Health aided the investigations by establishing branch laboratories at Hiroshima and Nagasaki for the ABCC. In 1975 the ABCC was replaced by a bi-national organization known as the Radiation Effects Research Foundation (RERF), with funding shared equally by Japan and the U.S.

A hundred thousand people in Japan, who had been exposed to various levels of radiation during the atomic bomb explosions at Hiroshima and Nagasaki, were followed for many decades to determine the health effects of the radiation exposure. This was crucial for understanding the health impact of medical equipment that uses radiation (such as X-rays and CT scans) and of radiation treatment for cancer. Such knowledge was also crucial for setting exposure standards for the higher background radiation that exists at high altitudes, such as for airline crew or for space missions. In order to gain such an understanding one had to know the actual radiation dose that had been received by the people in Japan who were being followed.

原爆放射線量再評価に関する 日米合同ワークショップ　昭和58年2月16日　（於 長崎）

Group photo of the members of the US-Japan Joint Reassessment of Atomic Bomb Radiation Dosimetry in Hiroshima and Nagasaki, which was coordinated through the Radiation Effects Research Foundation (RERF). Note Robert Christy standing at front center, the only person with white pants. (RERF photo)

The study group of one hundred thousand Japanese people (from among those who had been exposed to radiation at Hiroshima or Nagasaki) was followed for many decades by the Japanese health system, and their health was compared to a control group of Japanese people who had not been exposed to radiation. Any differences in the cancer statistics between these groups could be attributed to the radiation exposure. However, in order to determine the health risk of the radiation dosage one must be able to determine the radiation dosage that these individuals received, which turned out to be an extremely difficult task. In fact, a reliable calculation was not achieved until the group's 2002 report.

The nuclear explosions had generated large numbers of neutrons and gamma rays. In the case of the Nagasaki bomb, these were emitted almost equally in all directions, since the plutonium-239 implosion bomb was roughly spherical. However, the Hiroshima uranium-235 bomb was long and narrow, falling at an angle of about 15 degrees from vertical. The shielding effect of the nose of the bomb meant that somewhat less radiation was emitted in the direction the bomb was pointing than in other directions. In addition, the Hiroshima bomb was the *only* one of that design ever detonated, and thus even the strength of its explosion was not well determined until the 2002 report.

The neutrons emitted in the explosions were *fast neutrons*, and the majority of those emitted in the downwards direction reached the ground, where they bounced around and were slowed down — slow neutrons are much more easily absorbed by nuclei than fast ones — until being absorbed by a nucleus, which usually emitted one or more gamma rays (and sometimes yielded a radioactive isotope). However, Robert eventually realized that neutrons travelling roughly horizontally were more likely to be absorbed by the nitrogen or oxygen nuclei of the air (emitting a gamma ray at that point); several miles from the explosion, most of the radiation reaching the ground was comprised of gamma rays.

The radiation exposure of someone on the ground also depended on whether they were shielded by solid material between them and the direction from which the radiation was coming. Someone standing in front of a window would receive a rather higher radiation dose than someone who was behind a wall; someone in the basement of one of the (rare) concrete buildings fairly near the explosion might receive less of a dose than someone further away who was out in the open. Calculation of the shielding effects of the walls was made possible largely because most of the houses in these cities were of similar wooden construction.

Finally, the radiation dosage of specific organs in the human body differs, depending on where the organ is in the body, i.e., how much flesh and/or bone is present to shield the organ from radiation. Thus calculations of how the radiation was transmitted within the human body were required.

Robert Joins the Project

In 1981, the Working Group on A-Bomb Dosimetry Reassessment was organized in the U.S., with Robert Christy appointed as its Chairman. In early 1983, arrangements were completed for the US-Japan joint research program for reassessment of A-bomb dosimetry, which was coordinated through the above-mentioned Radiation Effects Research Foundation (RERF).

Several publications and reports came out of this group. Robert contributed to much of this work, but his name actually appears only on the executive summary of 1986 ("US-Japan Joint Reassessment of Atomic Bomb Radiation Dosimetry in Hiroshima and Nagasaki: Final Report — Dosimetry System 1986"). Robert co-authored this document with Eizo Tajima, the Vice-Chairman of the Nuclear Safety Committee in Japan.

Robert was also deeply involved in the theoretical calculations of how the neutrons and gamma rays from the atom bombs were transmitted, scattered, and absorbed. It was possible, but very difficult, to get measurements of how much radiation there had been at certain points on the ground, by looking at the trace amounts of radioactive isotopes that had been created by the neutrons in particular materials. And Robert realized that there was also a way to measure the gamma ray dosage, by looking at the effects on certain types of tiny quartz crystals in clay. These measurements allowed the group to test their calculations of how much radiation had reached the ground.

Robert made several trips to Japan in the early 1980's, as part of this dosimetry project. I joined him on one of these, in August 1983, the tenth anniversary of our marriage.

The Break-Up of the U.S.–Japan Collaboration

The bi-national collaboration was terminated in 1985 because of conflicts between the U.S. and Japanese scientists. The split occurred because the U.S. and Japanese scientists came up with different measurements of the radiation received by site samples from Hiroshima and Nagasaki. The U.S. measurements agreed with the theoretical predictions (primarily by Robert Christy, with theoretical backing from Dean Kaul), but the Japanese measurements did not.

The U.S. suggested that the Japanese measurements were faulty, but the Japanese were unwilling to consider this, so the collaboration between the U.S. and Japan broke off. In the long run, by 2002, it was shown that Robert's theoretical calculations and the U.S. measurements had been correct. Robert had always been outstanding in the reliability of his theoretical computations.

Separate efforts continued in the U.S. and in Japan, and Robert continued with the U.S. effort as the Chairman of the "National Academy of Science Committee on Radiation Effects Research Foundation Dosimetry." During the initial collaboration (1980 to 1986), the study's medical expertise had been provided by the Japanese. The work performed in the U.S. from 1986 to 1993 involved calculations and measurements in non-medical fields; but in 1993 a biologist and expert in radiation effects on animals and people, Bob Young, was brought in from the Department of Defense to co-chair the task with Robert Christy.

The American group that remained after the US-Japan collaboration broke up; Robert is at back center (Caltech photo)

The collaboration with the Japanese was resumed in 1997. By this time Robert was in his 80's, but he continued to contribute as a member of the working committee, though no longer as its Chairman. In 2002, an updated summary of the results was presented as "Conclusions and Recommendations of the US Working Group on the Reassessment of A-bomb Dosimetry."

The re-established post-1997 US-Japan dosimetry collaboration meets at Caltech; Robert is in the middle-left of the back row, just above the corner of the pool (Caltech photo)

Robert's Description of the Project

In his 2006 interview with Sara Lippincott, Robert recalled:

> I was invited to join a working group that was trying to understand what happened in the explosion of the bombs at Hiroshima and at Nagasaki. The main question we were trying to understand was: What were the actual intensities of gamma rays and neutrons to which the populations of these cities had been exposed? And, of course, the relevant factor was not the exposure of those who died almost instantly or within a few weeks. The relevant factor was the exposure of those who survived and have continued to live there.

The question I was involved in was: Ever since the bombs were dropped — or shortly afterward; I think it started in 1950 — a group called the Atomic Bomb Casualty Commission, formed first by the United States, started to gather data on the Japanese populations of these two cities, as to where people were at the time the bombs went off, and what were their symptoms. Did they get sick? And it tried to follow their health. Roughly once every year or two, these people come in for a check-up. And the commission has followed the health of about 100,000 Japanese ever since 1950. Some of these people had fairly high exposures to ionizing radiation, and some of them had physical problems as a result. Some have died of cancer, and others will probably die of cancer. Some had, perhaps, other physical problems. There were some women who were pregnant. One of the questions was: If a pregnant woman was exposed to this radiation, what would the effect be on her unborn child? So there are many things that medicine wanted to know about the effects of ionizing radiation on human beings. The number of people exposed to sizable amounts of radiation there gave doctors an opportunity to learn things that there was no other way to find out. When you can study 100,000 people who have been given some reasonably well-defined exposure to radiation, and follow all of their medical histories, this is very important information — not only for us and the Japanese but for the world.

But the one piece of information that was not very well known was: What was the actual radiation that these people had been exposed to? They had not been carrying meters with them at the time, because they hadn't been expecting the attack. So the purpose of this committee was to study the physical dosimetry. Once the committee knew where a person was and what the circumstances were — if the person was in a house, where was the house located? — it could calculate what that person's exposure to radiation was. I was part of the group of physicists — most of us associated with Livermore, Los Alamos, Oak Ridge, and some associated with Science Applications International — who were working on this, trying to find out what the exposures were.

This started in the early eighties, but I came into it around 1983. There was Bill Loewy at Livermore, George Kerr at Oak Ridge, Paul Whalen at Los Alamos — not famous, but busy workers in the trenches, who were very good.

We visited Hiroshima and Nagasaki. Mostly we had meetings, and they did work for which they got paid by the Department of Energy. We did discover certain physical artifacts that could be obtained from Japan and studied. The point was to find things that would not have been very much disturbed. For example, it was found that techniques that had been developed since the war for studying archaeology — for determining the age of pottery — these techniques were developed. And it was found that radiation on certain materials, like quartz crystals, causes excitations. And if the stuff stays cold, these electrons stay excited. Then, a thousand years later, you can warm up these little quartz crystals and look at them with a photometer to measure the light, and you find flashes of light coming out. It's called thermo-luminescent dosimetry. As I say, it was a technique pursued to study ancient pottery.

Roof tiles are made of pottery. And some were collected from old buildings that had not been totally destroyed, and taken to laboratories in Japan, England and this country, and they would grind up the tiles, get out the quartz crystals, heat them up, and look for the light and find out what the radiation was — the total number of gamma rays that had hit that roof tile. It was fascinating! We were able to encourage the use of these techniques and help assign people to do this work — to get the materials at all sorts of distances. Not too close in to ground zero, but at relevant distances from, oh, about a half a mile or a little less, on up.

Previous work had looked at radioactivity left in cobalt — which was looking for neutrons. Gamma-ray radiation had mostly been done by calculation, because [gamma rays] didn't stay around. But this work did quite accurately verify the very detailed calculations we had made. So there is a very good correspondence between the measurements and the calculations. The calculations could not really be trusted without having some measurements to compare them with, so the measurements were very important.

The other part of the story — the neutrons — we had more trouble with. And in fact, the neutrons are still giving us trouble. We have not yet found an adequate agreement between the calculated neutron intensity and the observed neutron intensity. Whether this is the fault of our calculations or observations, we don't know. But it's probably in the calculations. But the observations are good for telling you if you're on the right track. They don't tell you everything you want to know, but if you can pin down some things by measurement, then you can use your calculated model to tell you other things.

There were a few ways [to measure neutrons]. One was radioactivity — things that were irradiated at the time and still had a little bit of radioactivity left — not a meaningful amount but an amount that with the greatest of difficulty could be measured. And there was some radioactive cobalt left in iron that had been exposed. There was some radioactive europium left in some rocks and granite tombstones and things that people kept track of. These measurements, by the way, agreed pretty well with calculations on the Nagasaki bomb. They did not seem to agree very well at Hiroshima.

Then we discovered a new method — accelerator mass spectrometry. And that is, essentially, to look for unusual isotopes that are left over. The neutron is absorbed by a nucleus. If it's an unstable nucleus, it's radioactive. If it's nearly stable, it sits there. It doesn't give off any radiation; but mass-wise it is one greater. So you can then study how many of these nuclei there are around — and thereby measure the number of neutrons. And that method also did not seem to agree with the Hiroshima calculations. It did agree in Nagasaki. We're still trying to pursue the reason for that disagreement at Hiroshima.

There were more neutrons than expected at Hiroshima. It probably means that we didn't do something right with our understanding of how the bomb exploded. That was the gun-type bomb, which has caused us no end of trouble.

Anyway, that work is still going on. It would have gone faster, except that each time a fair amount of progress was made, and a few million dollars were spent, the government said, 'Well, gosh, we can't really afford to continue this,' so the work stopped, we ran out of money. And then some new ideas were developed, and there was new enthusiasm on the part of the DOE, or someone, to put a few more million dollars into it, and it was pursued. And now [in 2006] we're running out of money again. This is roughly the third time we've run out of money. We may be on the verge of understanding it, but we don't really have it properly worked out. It's a continuing problem. But nevertheless, we have already improved our understanding of what people were exposed to there. Of those 100,000 people, quite a number have died of cancer, but normally if 100,000 people die, roughly a third or a quarter of them will die of cancer — that's just the normal course of events. Now, perhaps only 40,000 of the 100,000 have died already. And a fair number of those have died of cancer. By comparing the number of cancer deaths among people who were very distant from the bomb to the number of cancer deaths among those who were close in, you can begin to find how many so-called excess cancer deaths are associated with the radiation. And it looks as though there will have been something like 5,000 extra people dying because of the effects of the radiation. That's a large number. But here you have a bombing in which roughly 100,000 people died as an immediate consequence — either in the explosion or from fire or from burns — within the first few weeks. So that 100,000 people died more or less immediately, and if 5,000 died in the next fifty years or so, that's a fairly small number. So the residual effects have been fairly small compared to the immediate effects.

What that tells me is, if there's a nuclear bomb and you survive the first weeks and months, then you've essentially made it. But the work of the committee is one of the principal sources of information for medical experts around the world on the effects of radiation on people. So it's been very important to try to get these numbers right.

Chapter 17

The Spring Valley Ranch: Fulfilling a Dream

Robert had always immensely enjoyed physical activity in the outdoors. When he was a student, his favorite summer job was one where he was out in the wilderness of British Columbia (in Canada), surveying all by himself. When he was in Chicago, he very much enjoyed swimming with Fermi in Lake Michigan. At Los Alamos, he enjoyed hiking with the other physicists, horseback riding with Oppenheimer, and cross-country skiing with Hans Bethe. At Caltech, he played tennis with Harold Brown and other top tennis players there, such as Jack Roberts and Harry Gray; as he approached his 60's was still defeating young men in their 20's. While living in Pasadena with his first family he hiked with his sons Ted and Peter, and took them skiing and on trips to national parks in western U.S. and Canada. He would enjoy swimming, snorkeling, and body-surfing in the ocean at La Jolla whenever he could. He became quite an accomplished downhill skier, and once won a bronze medal for speed skiing in an amateur competition at Aspen. He went for whole weekends skiing cross-country with Barclay Kamb, overnighting in rental cottages in the wilderness.

He had a great fondness for the beautiful land around Los Alamos, with the strikingly clear skies, the green valleys, and the dramatic mesas. One of his favorite regions there was the Valles Caldera. At the age of 66, when his administrative work was terminated and he was having difficulty getting back into research, I persuaded him to do what he had been unable to do before: to purchase a 240-acre mountain valley at an elevation of 6000 feet that reminded him of the Valles Caldera near Los Alamos. There was nothing there except nature — 40 springs, in acres of beautiful meadows and forests. There were no man-made structures. That was his "retirement" project, to construct a house with whatever was needed to spend enjoyable times there. Unlike the Valles Caldera in New Mexico it was close to home, only about 90 minutes from Pasadena.

We called it the "Spring Valley Ranch" both because of all the springs there and because it was intended to bring a new springtime into Robert's life.

That it did. We purchased 65-horsepower bulldozer that was over thirty years old, using money that we had planned to spend on a trip to the mountains of Switzerland. Every Saturday Robert had tremendous fun sitting on the bulldozer and cutting a dirt road along a steep hillside into the valley. Sometimes this meant digging eight feet deep into the rocky hillside. He worked on it most Saturdays for five years: it was called "Robert's Road." He tested the drivability of this road by trying to drive along it in his beloved green sports car.

He had a great deal of fun with this antique bulldozer and his sports car. For example, at the end of one road-testing day when he had to get both the bulldozer and car back down to the center of the valley, he decided that he didn't

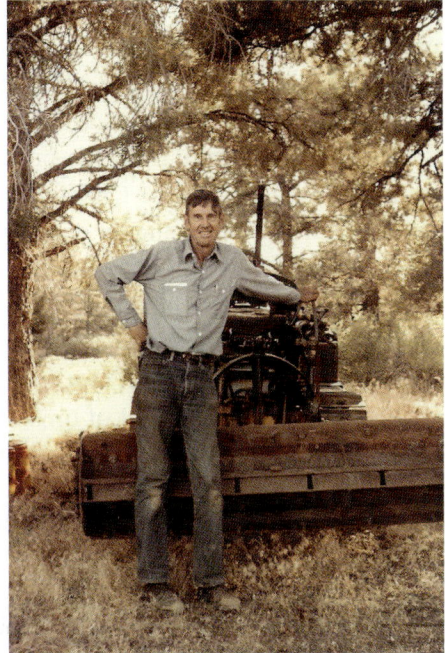

Robert with his 30-year-old bulldozer — still in good working order

want to deal with the hassle of walking back for the second vehicle, so he tried to come up with a plan to do it all in one trip. He tied the car to the back of the bulldozer with a rope in order to pull it along behind. The only problem was that he had to come down a hillside. Eventually the car began gaining speed and rammed into the back of the bulldozer, smashing the car's left fender. Not to be deterred, at a later date he decided to try his luck again and had another go at the rope tow strategy. History repeated itself; this time he took out the right fender. Not wanting to be thwarted, he changed tactics and attempted

Robert cuts a dirt road with his bulldozer in a flat part of the valley

the stunt on a third occasion, this time pulling the car in reverse. The result was that he damaged the car yet again, this time destroying the rear bumper. I made it his birthday gift to get the car repaired.

When Dick Feynman heard about this road construction he came out to visit Robert, who was sitting on his bulldozer covered with dust from head to toe. Feynman said, "Professor Christy, pleased to meet you like this." And Robert replied, "Dick, I'm having so much fun." Feynman proceeded to get a place for himself, a property in the mountains close to Pasadena that had a similar simplicity and beauty; but he never reached the stage of enjoying a bulldozer. He just used a wheelbarrow to get his supplies to the top of the hill where there was a small cottage with no road to it.

Robert spent a great deal of time surveying the property. Only one corner was established, and he used old equipment from Caltech's Department of Engineering to determine two of the other three corners. (The last corner still has not been established because of the mountainous terrain, with survey marks being too far away.) He also rented equipment to lay miles of underground water pipes from water sources to where the water would be needed.

Robert digging a trench for the water pipes

When he and the rest of the family stayed overnight, we slept on tarps, under the stars. Feynman and his family did the same when they came to visit and stayed overnight. The fact that there were coyotes, bears, and mountain lions in the area did not bother anyone. Many of Robert's other colleagues also came to visit. The land was properly christened with champagne, down at one of the springs' creeks, by Professor Bob Sharp of Geology and his wife Jean, and Professor Francis Clauser and his wife Catharine.

The Christy daughters were competitive horseback riders, and it was too costly to keep their horses in stables in Los Angeles. Therefore it was decided to place the older horses at the Spring Valley Ranch. Robert built several horse sheds to keep their hay dry. He would drive up every second day to give them two days worth of food.

Once, late in March 1991, there was a record snowfall up there at the ranch's 6000-foot elevation. Ten feet of snow fell in a few hours and packed itself down to five feet of dense snow. But the horses needed food, so Robert drove up in his sports car. He had to park three miles away from where the horses lived, as there was nobody there to plow the snow on the dirt road into the ranch. At the age of 75 Robert made himself a human snowplow, pushing his body forward step by step through the deep snow. Late

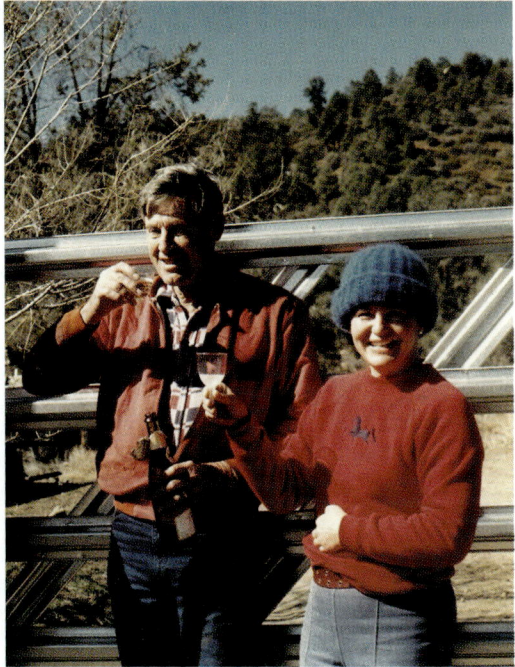

Robert and Juliana celebrating with champagne at the main entrance to the Spring Valley Ranch

Fall colors at the ranch

in the evening he finally reached the trailer that we had purchased for accommodations. When he finally got inside, dripping with sweat from his exertions, he turned on the propane stove to get some heat. I called every half hour through the night to ensure he was still O.K. The police offered to send in a helicopter to help him, but he replied that he was fine and didn't need any help.

The next morning he tried to get the hay to the horses, but kept stumbling over equipment that was hidden by the snow. Even the fences around the horses were completely invisible. The horses themselves had broken a path through the deep show to their water troughs. After he finished looking after the horses, Robert followed his own footsteps down the mountain, but found his car gone. It had been towed because it had not been parked completely off the road due to the snow banks at the roadside. Later that

Robert at the ranch

day he got the car and drove home, exhilarated that he had completed his task in spite of all the challenges.

In the following years Robert and I designed a comfortable house for ourselves and our guests, a big barn, a workshop, and beautiful and functional sheds for the horses. Robert designed many creative improvements for the house, such as an outdoor sprinkler system under the eaves in case there should be a forest fire, and hidden wiring for solar energy in case it should be installed one day. He designed an excellent gravity-flow water system for the structures. In case of an earthquake, the house was designed to withstand an acceleration of 4.5 g's, three times that required by law. He collected rocks from the nearby mountains to construct an impressive fireplace.

Robert had a horse at the ranch. It was 100% his, unlike the one he had shared 50-50 with a colleague at Los Alamos during the Manhattan Project (see Chapter 6). He was 75 years old when he obtained this horse, and he named it

"Blueberry Pie" because it had a rare blue roan color. When this horse died, Robert was 84 years old. He got a second horse, a spirited four-year-old colt called Cimarron (bred from a thoroughbred and a wild mustang caught in Nevada).

Robert had no qualms about jumping six-food-wide creeks on his horse Cimarron. Even at age 93 he was still riding his horse, in spite of having had major intestinal surgery two months earlier. When he and his family reached a clear area and he was asked

Robert and the rocks he collected to make the fireplace and the chimneys for the ranch house under construction

Robert and his horse "Blueberry Pie," the first horse that he owned from head to tail (unlike the horse from Oppenheimer that was shared between Robert and his colleague at Los Alamos)

whether he wished to walk or gallop, his answer was "Let's go!" And he was the one to lead the family in a gallop.

Robert loved the ranch and all the outdoor activities that it allowed. It had kept him young, just as the name "Spring Valley Ranch" implies.

Robert on his second horse Cimmeron and Juliana on her second horse Apollo, with daughters Ilia and Alexa in between

The ranch house that Robert and Juliana built

The living room of the ranch house

Freely roaming horses at the Spring Valley Ranch ranch

In one of the forests at the Spring Valley Ranch: Ilia on a "visiting" horse Willy, Alexa on Snoopy (which was actually Ilia's horse), Juliana on her first horse Shandu, and Robert on his first horse Blueberry Pie

Chapter 18

Robert's Health — Struggles and Successes

Robert was always very athletic and had excellent health throughout most of his life. He did have several bouts with pneumonia. He may have been more susceptible to it because he smoked for two decades early in his life, and was exposed to secondary smoke for another decade; in addition, he had had extensive exposure to diesel exhaust from the many hours he spent on his beloved bulldozer at our Spring Valley Ranch. He also had occasional back problems which he treated with simple back exercises. For the first eight decades of his life he encountered only fairly minor illnesses and injuries.

Robert was unusual among the men of his family in that he lived into his 90's. His father had died in a workplace accident at age 33 and both of his grandfathers had died young. His only sibling, his brother Jack, had died suddenly in his 40's. Robert was also fortunate in that he did not suffer from cancer or serious heart disease, Alzheimer's, arthritis, diabetes, hearing loss, or many of the other common ailments of advanced age.

Vision Problems

In the summer of 1991, when Robert was 75, we became aware that Robert was having some vision problems. At first they were relatively minor. We were starting construction on the main house at our Spring Valley Ranch, and when he looked at the footings of the foundation they didn't look straight to him. As the studs and walls were built he continued to see them as crooked, even though the contractors assured us they were straight. The problem was soon diagnosed as age-related macular degeneration (AMD), where the central cells of the retina (the region called the macula) stop their normal function and die off.

By 1997 he had lost the central vision in both eyes, but with magnification devices he was still able to read. He could no longer drive a car. However, even though his central vision was gone, he still had his peripheral vision and remained active. He was still able to travel by plane on his own and enjoy the outdoor recreational activities he had always loved.

Even though he was 81 years old, he was still galloping his horse across meadows and creeks on visits to the Hunewill Ranch. He was still able to ski at Mammoth and go on major trips — including the hiking trip in the mountains of Nepal that was described in Chapter 13.

Loss of a Kidney

Robert's first major injury happened in April 2005. He had developed a lung infection and had quite a cough. He got up in the night and started coughing repeatedly, and fainted and collapsed in the bathroom. Unbeknownst to him, during his fall he hit the toilet quite violently. When he woke up, he found himself on the floor, and decided to go back to bed. After lying in bed for a while he decided all was not well and called out to me. We decided that we should go to the hospital, even though it was the middle of the night. I helped him out of bed, but he collapsed on me. With him lying on top of me and now unable to get up, I had to wiggle out from underneath him. We laughed and laughed, not realizing the seriousness of the situation. It was the last laugh for a long time.

When I was finally able to emerge from beneath Robert, I saw that there was blood on the bed and floor where he had been lying, and on both him and myself. I called his physician and then called 911 to get help in bringing Robert to the hospital.

In the Emergency Department many tests were run. It turned out that during his fall Robert had broken ribs and shattered one kidney, resulting in major internal bleeding. The bleeding was so severe that he had emergency surgery to remove the kidney and control the bleeding.

Since our daughter Ilia was a physician in her residency training, she was able to confer with the trauma surgeon and the emergency room physician about Robert's situation. Both Ilia and I emphasized to them that even though Robert was almost 89 years old, he was still active and quite fit for his age.

Ilia and her fiancé Chris Wakeham, our younger daughter Alexa, and I all sat on the floor outside the doors to the operating room and waited nervously. It was a long surgery and Robert required multiple blood transfusions. Controlling the bleeding was difficult. His sons Ted and Peter came down from northern California to see their father.

Robert was in intensive care for several days. His children Ilia, Alexa, Peter, and Ted attended him in the intensive care unit while I was suffering through the final stages of chemotherapy. Robert then had to recuperate for a month in the hospital, with Ilia and Alexa spending each day with him and trying to keep up his spirits.

Robert with his supplemental oxygen tube, which can be seen running up his neck to his ear and around it to his nose

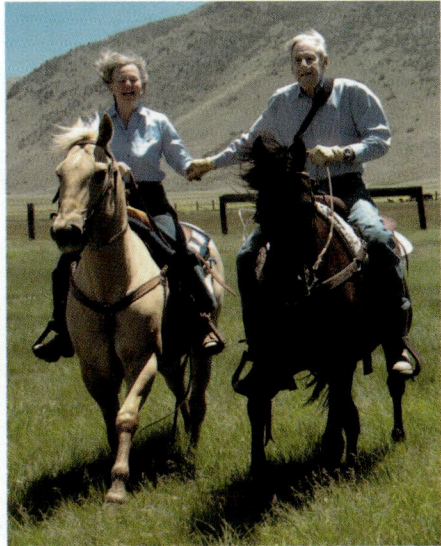

Ilia and her fiancé had planned to get married in October of that year. This proved to be a strong motivation for Robert in his rehabilitation. Despite his major accident, long hospitalization, and slow rehabilitation, he was able to walk down the aisle at Ilia's wedding and lead her on the dance floor. Soon he was also back on his horse again.

Robert (age 93) still galloping on horseback with Juliana at the Hunewill Ranch in 2009; the black strap of his oxygen concentrator can be seen running diagonally across his chest. He continued to ride until he was 94 years old

As Robert's friend Feynman had before, Robert now had to live his life with only a single kidney. He had to take powerful prescription drugs. He now suffered from orthostatic hypotension, with dangerously low blood pressure upon standing, that required careful management with medications and supplemental oxygen, especially at high altitudes. (The Hunewill Ranch and our own Spring Valley Ranch were both at 6000 feet.)

Even though he needed to carry an oxygen concentrator in a bag slung over his shoulder, Robert was still able to ride. The pictures on this and the previous page of Robert and me riding were taken in 2009, four years after his accident and surgery. He continued to ride until age 94.

In the summer of 2009 Robert travelled to Santa Fe, New Mexico to participate in an astrophysics conference on variable stars. He had been an early pioneer in

Robert (age 93) riding at the Hunewill Ranch in 2009. His oxygen concentrator is on his back: one can see the black sling it hangs from, and the clear plastic oxygen tube running from under his arm horizontally half way across his chest and then up to his neck (there the tube bifurcates and runs behind each ear, the two branches meeting at his nose; but that is not so easy to see in this photo)

this field in the 1960's (see chapter 10). His research had been fundamental, and his presence was an inspiration for many of the younger astrophysicists. While there he also gave an excellent summary of his life in science. He took a side trip from Santa Fe to Los Alamos he was still able to see all the key sights there. He recounted their significance, and sometimes added details to the stories told by the guides of the Los Alamos museums.

Intestinal Surgery for Adhesions

After Robert's accident and kidney surgery, he developed scar tissue in his abdomen. In the following years he had multiple intestinal blockages. In 2009 one of these required surgery to remove the adhesions (scar tissue) in his

CALIFORNIA INSTITUTE OF TECHNOLOGY
Pasadena, California 91125

Caltech

Jean-Lou Chameau
President

October 8, 2009

Dr. Robert F. Christy
Institute Professor of Theoretical
  Physics, Emeritus
California Institute of Technology
Pasadena, CA 91125

Dear Bob,

Ilia keeps your friends informed of your progress. I was pleased to hear that you are feeling stronger each day. Having spent time in a hospital a few years back, I appreciate how difficult the recovery from surgery can be. You are in my thoughts.

Your Caltech colleagues and friends want to see you back at the Athenaeum in the near future. We miss you!

Carol and I wish you the best in these trying times.

Best regards,

Jean-Lou Chameau

A "get well" note from Caltech's President Jean-Lou Chameau during Robert's 2009 hospitalization after his intestinal surgery

Robert about to set out on a ride with the family at our Spring Valley Ranch in November 2009, two months after his intestinal surgery

abdomen and remove a section of large intestine that had scarring and was contributing to his intestinal obstruction.

Each hospitalization and surgery was another setback for Robert, requiring extensive rehabilitation so he could regain his strength and walk again. But Robert was riding again two months after the intestinal surgery.

A Sudden Loss of Vision… and a Recovery

For many years Robert had been able to function relatively well with his peripheral vision. Magnifying equipment allowed him to read and write, and he could see well enough to walk by himself. However, his vision slowly deteriorated to the point that his reading became very difficult even with his magnifier — essentially he had to read one word at a time.

In the summer of 2010 a new decline in Robert's vision occurred. Suddenly he was almost totally blind. He could not see his hand, nor could he find doorways. This was due to blood leakage from his retinal blood vessels that resulted in the interior of his eyes (the vitreous humor) becoming almost totally opaque. He fell several times because he could no longer see where he was going but miraculously never broke any bones. He had been so athletic all his life that his body was still resilient enough to survive these falls without significant damage. From July to October 2010 when his vision was almost absent he did not move much. He lost more of his physical fitness than he realized.

In October 2010 he had eye surgery, which was successful. His vision improved a little and he was able to be more active again. Unfortunately, two weeks after the eye surgery when Robert hurried to answer the front door after hearing the doorbell, he fell and fractured a vertebra in his spine. This was very painful and required a surgical procedure to stabilize the fracture. Again he was bedridden and again he faced a slow rehabilitation to be able to walk once more.

On New Year's Eve, happy to be able to walk again at his usual swift and energetic pace, he tripped on the threshold where the original house connected to an added room and fell again. He fractured another vertebra, requiring bed rest. Another hospitalization followed, due to an intestinal obstruction that probably resulted from not being able to move about enough.

Through all of this, it was remarkable that Robert never let a single complaint pass his lips. He always bore his pain and discomforts stoically and did not allow himself to be upset by any necessary indignities. He always remained a gentleman. When asked how he felt, his normal answer was, "I'm fine."

During the last 15 years of his life Robert benefited greatly from unique contributions from myself, Ilia, and Alexa. I spent many hours researching macular degeneration and found novel therapies and treatments for a progressive disease for which conventional medicine had no effective treatment at the time.

Robert and Juliana in the summer of 2010

Even though Robert had supported my use of unconventional treatments in the fight against the cancer that had struck me, he was unwilling to try unconventional treatments himself. He always asked for peer-reviewed articles describing clinical trials. I would answer that these would come later, that we lived too early.

There was one exception. In August 2012 he agreed to umbilical stem cell injections in the hope that some re-growth of retinal cells might result. And indeed after about a month he reported moments of improved peripheral vision, and even occasional sharp vision in the center of his field of view.

Ilia applied her training and knowledge in internal medicine, often leading his physicians to the correct diagnosis. For example, at one point she noticed green bile around Robert's nostrils and suspected an intestinal blockage — which was confirmed shortly thereafter. After his second fall she convinced Robert to seek hospitalization for his intense back pain, convincing him that he had a fractured vertebra and not just a strained muscle.

Alexa organized his medications and his many doctors' appointments, and carefully kept track of his health on a day to day basis. She cherished the many hours she spent with him as she accompanied him to his doctors' appointments. Together, the three of us spent many weeks with Robert during his multiple hospitalizations, trying to keep his spirits up, pushing him to be more active, and watching for mistakes that sometimes occurred (Ilia was usually the one who caught the mistakes).

His Last Year

In the summer of 2011, at age 95, Robert was ambulatory with the aid of a walker. He only needed oxygen at high altitudes and was able to accompany me on my trips to San Diego to visit our daughters.

Robert in 2011, with his health much improved

Robert greatly enjoyed having lunch with his colleagues at Caltech's Athaeneum, at the "Table of the Knights" with Jack Roberts, Francis Clauser, Rudy Marcus, Maarten Schmidt, Ahmed Zewail, Charlie Campbell, Marshall Cohen, Lee Silver, Joseph Kirschvink, Cary Sieh, Jack Richards, Derna Fahem, Ed Stolper, and David Baltimore. He very much enjoyed these luncheon discussions and kept abreast of the news in science, politics, and world affairs to be prepared for the spirited conversations and debates that took place.

Unfortunately Robert's vision continued to deteriorate, and with his loss of vision he became less mobile. He had trouble walking even with a walker and needed help throughout the day for his usual activities. Luckily we found an exceptional woman, Carmen Menjivar, who greatly aided in his care in our own home during his last year. In the most difficult circumstances she treated him with respect and dignity, giving him warmth and kindness. He very much enjoyed living at home with me where he was still the "king of the castle."

I found a therapist, Will Locklear, who helped Robert exercise in the hopes of regaining muscular strength and avoiding muscular atrophy. Robert's strength was improving and he was able to get up from his bed without help. Will asked me whether Robert had any running shoes, to start him walking again now that he could get up on his own. Several times over the past years Robert had been able to learn to walk again after being hospitalized and lying in bed for lengthy periods. Our hopes were high.

In early 2012 Robert decided to avoid hospitalization for each illness by asking to be cared for in his own home by a hospice team. He had a wonderful nurse, Julie, who would come to the home to check on him. A guitarist called Rebecca came to the house once a week to sing and play familiar folk songs for him.

Robert and I enjoyed some wonderful days together. I would read to him most mornings for an hour or two from the letters that we had written to each other in 1970 and 1971, when we lived on different continents. He greatly enjoyed hearing these letters again. I was reading them to him chronologically and had reached the letters of February 1971. On special occasions such as his or my birthday, Valentine's day, or our wedding anniversary I would read to him a chapter from the draft version of this biography. At one point he smiled and said that he hadn't realized that he was living with his own biographer.

For our wedding anniversary on August 4, 2012 (which turned out to be our last one), I wrote him the poem below:

My wish list for Our Wedding Day
August the 4th 2012
39 years of Married Life to You

As you pass away from the present life
I wish I could pass with you.
You and I are the results of the laws of physics
That are the same everywhere and for all times,
On each of the many planets around us in the cosmos.
When I pass on I wish to meet you somewhere
With my wish list given now

To be born again with you
Together on the same planet
And in the same time period
And this time not 26 years apart
But of the same generation
To be your first wife this time.

And I wish not to make the same mistakes
That I made in the present life with you,
I would give more time to you
And not look after so many others
Who asked too much of me,
Whose needs burned both of our lives down,
And give more of my gift of energy to you.

I wish I would not have worked so hard
And expended so much energy on investments
To help earn a pension plan for us.

And I wish to have persuaded you
To get rid of our ranch
When a nasty neighbor sued us
With a frivolous all-consuming court case
That took four years of good life out of us.
And I wish I would not have undertaken
The enormous constructions at the ranch.

I wish not to have trusted the ophthalmologist
But searched for a winning solution myself
For your supposedly incurable retinal disease.
And I wish I had insisted that you listened to me more
Than to the "Come back, Professor" empty statements.

And I wish I would not have thought
That life is so long, with so much more to come;
Rather I wish I would have spent
More irreplaceable precious moments with you.

Since I do not believe in the destruction
Of the human spirit, the core of life,
I wish to be given the chance
To be born again in life with you
Without making the mistakes of this one
So that you can feel my hands again,
Hear the lilt of my voice and feel my warmth
As yours again.   –Juliana

On Sunday September 23, 2012 Robert was happy to experience Alexa fulfilling her dream of putting her newborn baby daughter Mila onto his lap — Mila was born September 1. Then a week later on Sunday September 30 he likewise enjoyed meeting Ilia's newborn son Kelso (born August 20). He was touched that Ilia's son's full name was Kelso Robert Christy, named after him. He was intrigued by the unusual name Kelso and wanted to know where it had come from. Ilia told him that there had been a famous race horse called Kelso in the 1950's. Robert recalled that there was also a unique sand dune formation in the Mojave Desert known as the "Kelso Dunes."

Robert in His Mid-90's: Planning to Write Scientific Papers

Despite his advancing age, Robert's mind was still active. In 2009 he began making plans to work on several scientific papers. For years an unfinished paper in nuclear physics had been lying on his desk. He was proud of this paper and very much wished to complete and publish it. He planned to do this in collaboration with Arnold Boothroyd, who had obtained his Ph.D. at Caltech in 1987 under my supervision and who had collaborated with me on many papers since then.

Robert was also planning a scientific collaboration with me for the first time, as well as with Arnold, on a variable star project. Arnold and I would use our computer program to determine the star's structure during its long-term evolution up through the point late in the star's life when it would be expected to become a variable star. However, since this computer program dealt only with the long-term evolution it could not be used to tell whether or not the star encountered variability.

Robert's computer program was still on punch cards in his office. A collaborative project took shape with Robert's son Peter taking Robert's old punch cards to an antique punch card reader (one of the few left still functional) to read Robert's old program subroutines into a computer. Arnold then began the task of updating this program into modern programming code and adding updated physics to it.

Robert's updated computer program would then use as input data the basic structure provided by our stellar evolutionary program's calculations, and determine whether the star was variable on relatively short timescales (hours to weeks) at that point in its lifetime. If so, his program would determine all the characteristics of the variability, which could be compared with observations.

As it turned out, the preparatory programming updates still had not been completed when the end came.

Passing On

Robert's passing was remarkable in two ways. First, he had predicted decades ago that he would die of a lung illness, and that is what happened. Second, his passing was unusually peaceful, very much as he would have wished.

Robert was coughing a great deal on the night of Monday October 1, 2012. As a precaution I put Robert on his supplemental oxygen. Despite the oxygen his coughing did not stop, but all his vital signs continued to be normal. However, on Tuesday morning as I measured his oxygen saturation it suddenly dropped from about 95% (normal) to 80% (which is dangerously low). While I repeated the measurement, it dropped to 70%, then 60%. I called for help. A wonderful home nurse, René, came immediately with medications to help him breathe. Ilia and Alexa set out as soon as they could from their San Diego homes, arriving in the early afternoon with their newborns (the babies could not be left behind since they were being breast fed).

Rebecca — the musician who came on Tuesdays to play her guitar for Robert and sing the country songs that he loved — arrived as usual that afternoon. We gathered around Robert's bedside in the master bedroom. Rebecca played her guitar and led the singing while Ilia, Alexa, and I all sang along. Robert's breathing seemed to have improved and he sometimes tried to sing along too.

I stayed up with him that night till 3:00 a.m., together with a new night nurse. Robert seemed more at ease; there was no more coughing. On Wednesday

Sunday September 30, three days before Robert passed on

morning I spent some time at Caltech talking to the Development Department about the possibility of setting up an undergraduate physics scholarship in Robert's name. When I returned to Robert, I informed him of what I had been up to and that it seemed possible that we might be able to fund such a scholarship. He listened happily.

The nurse who was with us at midday on Wednesday was Julie, who had frequently visited since February and was familiar with Robert. I asked her what the prognosis was. She said that she could not give an answer: every patient behaved differently, and he might possibly recover. I knew of course that he had had similar breathing problems a number of times before and had always managed to recover from them. He was a survivor.

I sat down on a little stool to the right of his bed and put my small hand into his large folded hands, as he had always enjoyed since our first days together in Portugal in 1970. I kissed him on his right cheek and then his forehead. At the same time Ilia, who was sitting on a little stool at his left holding her new-born son, kissed him on his left cheek. Robert was smiling. Alexa entered the room holding her new-born daughter. For a second time I kissed Robert's cheek and forehead, not letting go of his hands. The nurse beside me gave me a strange look, and I asked her why. She told me that Robert had passed on. There was no heartbeat. It was 12:45 on Wednesday October 3, 2012.

I couldn't believe it. My lips had been very close to his, but I had sensed nothing. There had been no struggle for air, no breathing problems, nothing to indicate that he was passing. But both babies had started to cry softly as if they had sensed what I had not. I went to gather my wedding poems and put them on his chest, then proceeded to the rose garden to get the orange-yellow roses whose color he had always liked so much. I telephoned Rebecca, and she kindly joined us again. As on the previous day, we gathered around Robert's bedside and sang the songs he had loved — but this time he was quiet and still. I called a chaplain, who came and helped us by reading poems about passing on that he had written himself. Later Carmen Menjivar dressed Robert as she had done many times during the past year, but this time I provided her with his wedding suit.

Heidi Mason, the wife of Robert's son Peter, arrived within an hour of Robert's passing — Peter had asked her to come on his behalf. Alexa's husband Mani Jodat and his mother Mina Jodat also arrived to be with us soon after Robert's passing.

Robert looked so handsome and so young: the wrinkles in his face seemed to have smoothed out. He was smiling peacefully, but he was so still. I could not let him go. I decided to keep him for another night and another day in our master bedroom (while I slept in the garden study); but since temperatures were unusually high I had to let him go to a cooler place the next day at 3:00 p.m. When they took him they covered his lower body with a lovely regal-looking red blanket.

Robert's passing was exactly as he would have wished. In his first letter to me, on September 13, 1970 while he was on his flight from Lisbon to Los Angeles, he had written,

> I can still feel your hands, see your face, and hear your voice. I hope I always can. It was a wonderful, wonderful thing to happen to your Robert

Robert's Final Resting Place

Robert was buried in the Mountain View Cemetery in Altadena, close to his friend Richard Feynman and next to the gravesite where Rudy Marcus had buried his wife Laura. Professor Marcus, who is another brilliant professor at Caltech, had been Robert's frequent lunch-partner at the Athenaeum, Caltech's faculty club.

There was a beautiful ceremony in the cemetery's chapel on Saturday October 20 that I had organized in the two-week interval. Unfortunately I could not be there myself because of an emergency retinal surgery on October 18; Ilia presented my eulogy. I had arranged for 96 New Freedom roses to cover Robert's coffin, one for each year of his life. A gifted soprano, Isabel Zepeda, sang "Ave Maria" as had been done at our wedding. Stephen Grimm, a deep-voiced musician, sang Schubert's "Das Litanei" and several folk songs that Robert had loved. Eulogies were given by our nephew Craig Cameron, our daughters Alexa and Ilia, the Chaplain, Robert's colleagues Professors Jack Roberts, Francis Clauser, Lee Silver, and Mrs. Marge Lauritsen Leighton. Close relatives and friends volunteered as pall-bearers: Ted Christy, Alex Levy, Mani Jodat, Saul Baroccio, Craig Cameron, and Erich Neumann. Robert was finally carried to the gravesite in a horse-drawn hearse.

Robert's final resting place at the foothills of the mountains to the north of Caltech is fitting for him. He had always had a special love for mountains.

Chapter 19

An Innate Grace

The Essence of the Man

Robert Christy was a daredevil. He was gutsy. For example, when he was a little boy he would climb up the huge pine trees in Vancouver and slide down from branch to branch until he reached the ground. After his father died he helped earn a little income delivering newspapers by bicycle, speeding up and down the steep hills of his neighborhood and throwing the newspapers to the doorsteps with a precise aim without slowing down. He would sometimes grab onto streetcars to pull him and the bike up the streets. As a teenager, his favorite job was surveying in the mountainous wilderness of the interior of British Columbia, alone in the mountains among the grizzly bears. When he was at Los Alamos he would enjoy the powerful thunderstorms and lightning, saying, "It's part of nature." Later in Pasadena, when earthquakes struck he would say, "Let's enjoy it; let the earthquakes rock us in our bed." When we bought our ranch land there was no house there, and for many years we enjoyed sleeping under the trees in sleeping bags. It didn't matter if there were bears or the odd mountain lion around. When we finally were able to build a ranch house there, he designed the living room roof such that we could hear the pounding of the rain on it.

When our first daughter Ilia was born, Robert got a sports car (his beloved green Datsun 280Z) and a trailer to hold the essentials for a baby. Robert drove that sports car for over two hundred thousand miles. He always enjoyed overtaking everyone ahead of him — speeding tickets were quite frequent. He could easily have become a race car driver.

Robert had an infectious joie-de-vivre. He loved life, living it fully. His early life was full of misfortunes but he did not let this get him down. There were so many small things in daily life that brought him joy. Sunshine in the sky was something that he would always notice and enjoy. He was enthusiastic about flowers and trees.

He especially loved the mountains and the desert — he wrote to me in one of his first letters that I should know that he loved the mountains the most of all things, after me. He loved skiing fast downhill: his favorite photo was one of him at Aspen where expert skiers had set up a course, and he was photographed on it during his run in which he won a bronze pin in that competition. When he was working at the University of Chicago in the 1940's he used to swim regularly with Enrico Fermi in Lake Michigan. He learned to ride a horse that Oppenheimer gave jointly to him and a colleague at Los Alamos: they would do physics on horseback. He enjoyed cross-country skiing, which he had first learned at Los Alamos from Hans Bethe. He enjoyed playing tennis at Caltech with Harold Brown and the other top tennis players among the faculty and graduate students. He loved to play tennis and to beat good tennis players. When Robert and I met at a scientific meeting in Portugal, he was overjoyed to see that he could teach me to ride the big waves, and that I enjoyed it as much as he did.

Robert loved to ride horses. He took polo classes at the equestrian center in Los Angeles when our daughters played polo at Stanford University — he was about eighty years old then. He actually prepared the ground for a polo field at our ranch. Every year we would spend at least one week at the Hunewill dude ranch behind Yosemite National Park, where there are over a hundred horses to pick from to ride. Robert and I would ride there every day, galloping through the meadows while holding hands. He used to ride a horse called Rocket, jumping him over creeks up to six feet wide while in his seventies and eighties. He was still riding and galloping with me at age ninety-three. When we rode at our own ranch through the forest there, even though his vision was very poor, he would say, "Oh, my horse can see." When we came to an open area he would say, "Let's go," meaning that he wanted to gallop. Then he would lead our family group in doing so: him, myself, and our two daughters Ilia and Alexa, who are also excellent riders. Very rarely did Robert fall from his horse... and always with grace and good humor.

He bought his first entire horse when he was in his seventies: he named the horse Blueberry Pie. This was a versatile and stately horse, a good fit for Robert. When this horse died, Robert got his last horse, called Cimarron, whose ancestors had been wild horses roaming in Nevada. Robert was 84; the spirited colt was 4. He rode Cimarron until he was ninety-three. When he had major intestinal surgery in his late eighties, it was only two and half months before he was riding this horse again.

Robert had a subtle sense of humor. Harold Brown, who had been President of Caltech while Robert was Provost, wrote that, "Those of us who were close to him also enjoyed his quiet sense of humor." In one of Robert's files I found an unattributed description saying,

> In Bob's typical nonchalant manner he recounts his wartime work at Los Alamos, New Mexico on the critical assembly for the plutonium bomb, also known as "The Christy Bomb." Bob states that his height came in handy when he needed to make adjustments on the bomb.

In a letter to me written in December 1970, Robert said:

> I got your camera brochures today. … Until I know whether or not I will really use something, I do not want to buy an expensive one. In fact, until then, I have no adequate basis of experience to pick out a good product. So I am inclined to purchase something of intermediate price and when I find out what I really want — then I know enough to spend more and get what I want. This is what I plan to do about the camera. If I get the bug then, with your help, we can investigate the kind we really want. (Do you think that, having been married once I am now in a better position to find a first quality wife for the second time?) Remember I have always said my jokes are not always recognized. But they are not always said totally in jest either. In that respect I think I have found it but still haven't completed the negotiation (for the wife of course).

Robert loved to travel to unusual places. One of his favorite trips was across China from the eastern coast to the Himalayan Mountains and the Karakorum Highway in Pakistan (not far from Afghanistan). I also remember a trip to Bangkok in Thailand, where we had fun doing all sorts of mischievous things. When he lost most of his vision and I had lost the central ligament in my knee (due to a riding accident), "the blind and the lame" went hiking in the Annapurna Mountains in Nepal in order to raise our spirits. He eagerly trekked up and down rocky paths through the mountains, relishing the adventure and appreciating the beauty of the mountains. He always enjoyed being in the mountains.

Robert was unbelievably big-hearted and generous. Most people who wanted financial support, be it private or political, got it from him. When he believed in something he gave it all he had, frequently sacrificing his own personal wishes and desires in order to do so.

When it came to medical problems, he and I had opposite opinions on how to attempt to cure them. When cancer struck me in 2003, and it was discovered to already be in my lymph system, he was my strongest supporter. He let me handle it my way, despite the large monetary expense of doing so and even though he didn't believe in my techniques. I learned to support him even though I didn't believe in his techniques. He always asked for peer-reviewed articles (like in physics) when I suggested a new medical technique to him. I would tell him that we lived too early: they would come later.

Dick Feynman used to park in front of our house on Arden Road next to Caltech, and we would sometimes walk the remaining few steps together. He asked once, "What is it between Robert and you, that you still have that magic spark flying between the two of you?" I answered, "He gave me space, he allowed me to continue to grow."

Robert's Philosophy on Publications

Unlike many eminent scientists, but like Feynman, Robert Christy never exhibited a need for recognition. He carried out his tasks because of the inner satisfaction that it gave him. When his work did result in major accomplishments he was appealingly modest about these achievements.

Scientists are typically judged by the number of peer-reviewed articles they have published, and on the usefulness of any books they have written. However, Robert did not choose to follow this usual pattern. He says, "I never believed in publishing quantity rather than quality." He wanted only his major contributions to be associated with his name.

He was Oppenheimer's chosen successor at Caltech — Oppenheimer called him "one of the best in the world" — but Robert never bothered to put together a list of his own publications. At one point in 2001 his secretary, Helen Tuck, tried to put together a list from the 28 reprints that Robert found in his files at Caltech; but since he was already legally blind at that time, this attempt was probably incomplete. In a letter of February 7, 1961 Neil Tanner of Clarendon Laboratory in Oxford joked, "It seems that I am almost as slow at writing letters as you are at writing papers." Colleagues have stated that Robert opened up whole fields with preprints that he never bothered to publish, because he felt that they were not important enough to be worth the trouble.

Graciousness with Finances

Even before he was ten years old Robert earned money as a child delivering newspapers on a bicycle, and on Saturday afternoons as a caddy on a golf course;

but he chose to give all of his earnings to help his widowed mother pay the household expenses. Later, when he turned 21 (in the middle of the Great Depression), he received a large cash inheritance from his paternal grandmother in England — enough to buy several houses or support him for many years. He volunteered to give all of this cash to his older brother, who was supporting their maternal relatives in Vancouver — even though at that time Robert's income from his fellowship at Berkeley was barely enough to survive on.

After the end of the Second World War Robert's first wife Dagmar learned that a cousin of hers in Hamburg, Germany had a four-year-old son, and needed financial help. At Dagmar's initiative, she and Robert took what they could out of their own modest income and sent monthly financial support to this cousin for many years.

About a decade later Robert's brother Jack died in his 40's due to a ruptured aorta, leaving behind his wife Jean Christy and their two young daughters Jane and Gail. Again Robert and his wife Dagmar decided to send monthly financial support to his brother's family, despite being on a tight budget themselves. These gifts too continued for years.

Another decade later, when Robert was in his 50's, Dagmar asked for a separation, followed later by a divorce. They divided their assets: she chose to take their savings, while Robert agreed to take their house. As it turned out, he paid her so much alimony that at the time of their separation her after-tax income would have been roughly twice his. They shared the same lawyer for the necessary legal procedures.

When Robert was finally able to pay to spruce up the house enough that it could be sold, he gave the majority of the net proceeds of the sale to his two sons, Ted and Peter, to help them get forward in their lives. This was a joint decision between him and myself.

When Dagmar died in 1982, it turned out that she had chosen to appoint Robert as the executor of her will, despite the fact that he was married to me at that point. Dagmar's will had to go through probate court proceedings, taking a fair amount of Robert's time. He was entitled to reimbursement from her estate for his time, and she had also granted him the right to take some of her household goods if he wished. Again he chose to take nothing for himself. Everything went to their two sons.

Thereafter, Robert looked after me — his new wife — and our two daughters with a similar commitment and passion.

With Colleagues and Family

Robert's modesty showed up already when he won the Governor-General's Gold Medal as the top high school student in the Canadian province of British Columbia. He would say that he only won it because they changed the format of the exams from essay to multiple choice. When other graduate students needed a team of two students to understand Oppenheimer's lectures (one listening and the other frantically taking notes), Robert did it all alone, and took it for granted. When he obtained instant fame at Los Alamos for calculating the critical mass of nuclear fuel needed for a "water boiler" reactor to within a few percent, he attributed this to luck. When he made a critical breakthrough for the design of the first atomic bomb, he said that many people had been working on it and he was inspired by their work. When he lacked the time to carry out his own physics research that he so very much enjoyed (and was so good at) because he had generously offered to carry the entire burden of graduate student supervision for both himself and Feynman, he merely stated that he wanted to have Feynman as his colleague, even though he deeply missed doing research himself.

He always considered the well-being of the group to be more important than his own personal goals. He was never driven by selfish motives. Feynman insisted on his sabbatical (a year of paid leave) in his very first year at Caltech because he had been due such a sabbatical at his previous university, Cornell. On the other hand, when in 1953 Robert had earned his own first sabbatical after seven years of teaching at Caltech, he did not follow his own desires but instead gave up his sabbatical so that his young sons would not have to change schools. In four decades at Caltech, he took only three sabbaticals. The first was for nine months (rather than the full year to which he was entitled), when he joined Oppenheimer in 1960 to study at the Institute for Advanced Study at Princeton. For his second sabbatical, he spent just a few months at Cambridge in England.

After he had served Caltech in administration for a decade as Provost (i.e., Vice-President for Academic Affairs) and for two of these years as Interim President of Caltech as well, Robert was again entitled to a sabbatical year. He was attempting to get back into research at that point in his life, and looked forward to studying the Earth's atmosphere in connection with meteorology and climate predictions. He had planned to use the newly-obtained satellite observations of ocean currents to help advance meteorological models of the atmosphere. However, in this year our little daughter Ilia was applying to private schools (since the local public school had too much crime and violence) but had been turned down due to the large number of applicants. To help her get admitted Robert again sacrificed his planned sabbatical year at the La Jolla Scripps Institute of Oceanography, taking it at Caltech instead. (It is interesting that not only did Ilia get admitted to a private school due to Robert staying in Pasadena, but today she is a physician at a Scripps Medical Clinic in San Diego near La Jolla.)

Robert loved physics, and particularly his own independent research. When he came up with a major breakthrough in the theory of pulsating stars, he again stated that he was only able to accomplish it due to his wartime work on shock waves associated with the detonation of atomic bombs, and that the others who had been working on the problem for a decade before him were simply not lucky enough to have had the experience of working on the Manhattan Project.

There were only a few research projects that Robert expressed himself as being especially proud of. One was his cosmic ray work, another was the determination of which nuclear reactions were primarily responsible for the Sun's energy generation. He was fondest of his work on variable stars, which he had carried out in the 1960's. However, he was also very proud of his crucial contributions to the dosimetry for Hiroshima and Nagasaki.

Robert had a large number of students, both in graduate courses and as thesis students whom he supervised. He had a tremendous impact on these students. Whenever he traveled, it seemed that somebody in a responsible position would greet him and say, "Do you remember, I was your student."

I was once on a flight coming back from an international meeting in Bulgaria. When a man sitting next to me learned that I came from Caltech, he enthused over someone from Caltech who had had the greatest impact on his life, and whom he admired more than anyone else. In the end I asked him the name of this person he so greatly admired, and he replied, "Robert F. Christy."

Like the greatest of men, Robert was humble and gracious about his accomplishments and willing to work for the good of others, not just for himself.

Index